HATCHED

HATCHED

Dispatches from the Backyard Chicken Movement

GINA G. WARREN

UNIVERSITY OF WASHINGTON PRESS
Seattle

Copyright © 2021 by the University of Washington Press

Design by Katrina Noble
Composed in Scala, typeface designed by Martin Majoor

25 24 23 22 21   5 4 3 2 1

Printed and bound in the United States of America

All rights reserved. No part of this publication may be reproduced or transmitted in any form or by any means, electronic or mechanical, including photocopy, recording, or any information storage or retrieval system, without permission in writing from the publisher.

UNIVERSITY OF WASHINGTON PRESS
uwapress.uw.edu

LIBRARY OF CONGRESS CATALOGING-IN-PUBLICATION DATA
Names: Warren, Gina G, author.
Title: Hatched : dispatches from the backyard chicken movement / Gina G. Warren.
Description: Seattle : University of Washington Press, [2021] | Includes bibliographical references. |
Identifiers: LCCN 2020045912 (print) | LCCN 2020045913 (ebook) | ISBN 9780295748627 (hardcover) | ISBN 9780295748634 (ebook)
Subjects: LCSH: Chickens. | Poultry. | Hens.
Classification: LCC SF487 .W377 2021 (print) | LCC SF487 (ebook) | DDC 636.5—dc23
LC record available at https://lccn.loc.gov/2020045912
LC ebook record available at https://lccn.loc.gov/2020045913

The paper used in this publication is acid free and meets the minimum requirements of American National Standard for Information Sciences—Permanence of Paper for Printed Library Materials, ANSI Z39.48–1984.∞

To my parents, Michael and Linda, with love

CONTENTS

Author's Note   ix

1. Flock Frenzy
The Rise of Backyard Chickens   3

2. Biking for the Birds
Silicon Valley's Tour de Coop   28

3. Urban Agriculture
When Chickens Come to Town   55

4. A Freegan Flock
Dumpster Diving and Limiting Waste   84

5. Pampered Poultry
Designer Chickens and the People Who Love Them   109

6. Fowl Feast
What Can and Does Go Wrong   132

7. Eating Bugs for the Environment
The Chickens and I Share a Meal   157

8. Productive Pets
The Rise of Broilers   172

9. Slaughterhouse in the Backyard
Culling Hens   191

10. Waste Not, Want Not
The Offal Truth   213

11. After Harvest
To Get to the Other Side   228

Bibliography   245

AUTHOR'S NOTE

There are a number of professional sources for chicken care, health, and public safety. This is not one of them. While this book is at times educational, it is not a how-to or a veterinary manual; it is representational of my personal experience, perspective, and (sometimes failed) experiments.

For practical resources concerning chicken ownership, health, and proper handling practices, contact your local veterinarian or the Humane Society. Although this is not an exhaustive list, you can find contact information for poultry veterinarians by state at poultrydvm.com.

I do not recommend using this as a guide for animal slaughter. If you are interested in this practice, I suggest contacting a small-scale farmer in your area who raises chickens or someone who is experienced with slaughtering animals. I do not recommend learning how to perform this process without the careful, hands-on guidance of someone who knows what they are doing.

For information about public health and diseases, consult your doctor, the Food and Drug Administration's website, or the Centers for Disease Control's website.

All of the individuals who appear in this book are identified by their own names, unless a pseudonym is otherwise specified. Certain individuals preferred to be identified only by their first name instead

AUTHOR'S NOTE

of their full name, accounting for differences in how individuals are recognized in this text.

Thank you to the editors of *Orion* and Terrain.org, who published portions of this work in a substantially different form under the titles "The Chicken Project" and "Eight Roosters," respectively.

HATCHED

Flock Frenzy

THE RISE OF BACKYARD CHICKENS

SOMETIMES MAKING ONE decision means making a thousand others.

It is 6:00 a.m. and I am squatting on the bathroom floor scratching chick feces out of the shag shower mat with my fingernails. I remind myself that I chose this. The four creatures responsible for the mess chirp wildly from a large plastic tub to my right. They are hungry. Even at a few weeks old, they associate my presence with food. I can't pick the poop free with my short nails, the chicks whine in unison like a sad boy band, the space has become their nursery rather than my bathroom, and I am not amused.

I remind myself that it is ungracious to be too frustrated. I chose this.

I chose to increase the overlapping territory in the Venn diagram between what I consume and what goods I can understand as part of a continuous process. It's almost impossible to imagine what actually goes into a jar of store-bought mayonnaise because it's just

CHAPTER 1

mayonnaise; something about its simplicity obscures what it actually is. It's easy to forget the eggs that go into the jar, let alone the chickens and the land that are represented by the existence of something so seemingly wholesome. I want to understand the implications of my choices with more depth: What are the particular requirements of what I eat? What kinds of resources, animal labor, and human participation are required when I eat meat and eggs? Is it possible, instead of trading money for food, to exchange bodily things like labor, time, and care?

Because of that one choice to try to occupy my role as consumer with a little more awareness, I signed up for a thousand others. Eventually it will mean constructing a chicken run in the backyard, biking around an unfamiliar city to see urban chickens, crawling through dumpsters until wet objects weasel into the open lip of my waterproof boots, and biting into the soft pads of stewed chicken feet, but for now, in this moment, it just means finger-tweezing wet crap from the fuzzy blue fibers of a bath mat before I've eaten breakfast or had coffee or even used the toilet in a bathroom that is not my own—at least not anymore.

My desire to get hands-on with my food isn't unusual. In some ways it's just a reflection of a countercultural zeitgeist that's gone mainstream. Over the last decade or so organic agriculture, ethical meat production, and sustainable farming have risen to the forefront of society's consciousness. A trip to the grocery store reveals pasture-raised, cage-free, free-range, and grass-fed foods. People can be locavores, eat slow food, and dine at farm-to-table restaurants. Even corporations are paying attention: the conglomerate General Mills owns the organic brand Annie's Homegrown. Kellogg's owns Kashi. The list goes on, but behind all the buzzwords is the idea that food systems can be reenvisioned and the sense that our conventional models of production and consumption are not serving society or the environment. There seems to be a hunger for something more nutritious and understandable than anything that's plastic-wrapped and marinated in preservatives.

On the tide of this hunger arrived a swell of urban chickens. People all over the United States are enacting back-to-the-land, sweat-of-your-brow-style virtues and building or buying coops for their neighborhood backyards. Although chickens raised in residential lots are often kept, at least in part, for their productive abilities, they're frequently treated more like pets than livestock, and the people who own them aren't always farmers. Some backyard chickens sleep inside, wear diapers, learn to pick the queen of spades from a pack of cards, or go swimming at the beach. Although chickens in urban environments might have once seemed like idyllic throwbacks to country life, they've been adopted by celebrities like Kylie Jenner and Jennifer Garner.

Chickens are a lot more mainstream than veganism and a little bit like kombucha: super weird twenty years ago, now somewhat popular and made even more so by logos, brands, and hashtags. Chickens provide a heavy dose of pastoral whimsy to even the most concrete of jungles, and they grant consumers access to local, fresh eggs overwhelmingly believed to be more nutritious, safer to consume, and better tasting than the store-bought alternative. Today the backyard chicken industry churns out both practical items—like organic feed, prefabricated coops, and waterers—and arguably impractical ones—like chicken hats, leashes, and designer carriers. Through reenacting old ideals of simpler, more self-sufficient bygone times in newly imagined ways, backyard chickens are reflective of an ambiguity growing within individuals about environmental sustainability in contemporary life, but they are still entrenched enough

CHAPTER 1

in capitalism to provide a market for hen tutus made of pastel taffeta and periwinkle tulle.

It's impossible to tell the story of backyard chickens without telling the stories of the people who care for them. Although cleaning the bath mat I had no idea of it yet, asking questions about backyard chickens as a phenomenon will lead me to people who grow their own mealworms for hens, who let their chickens live inside, who take their chickens to the beach, who go on road trips with their chickens, and who bring them to private air hangars. It will lead me to activists and rescuers and dumpster divers and farmers, but most importantly, it will show me how people's ideas about production, consumption, and agribusiness are changing and how chickens in a backyard can be indicative of shifting desires, priorities, and ideas about food. The concept of an ethical omnivore is a fraught one, full of complicated questions about what we should eat, how we should eat, and whether or not it's possible to consume an animal that didn't want to die and still retain any claim to morality. But this, too, is central to some of the questions raised by keeping chickens as productive pets.

The subversive character of people who imagine a future in which eggs come from backyards and not supermarkets, chickens are rescued from factory farms, and trash becomes food is alive now in the backyard chicken movement. There's also something slightly strange about anyone who finds a chicken charming. Chickens are friendly, social creatures, but they are also quirky at best. Given the right light, their beaks look menacing. Their legs are scaled. For most people living in the United States, we have spent our lives imagining chickens as creatures who generate store-bought products rather than friends in our gardens.

Chickens have become emblematic of a new brand of urbanism that reimagines what a city might look like and challenges our addiction to destructive consumption. People usually get chickens because they decide, for one reason or another, that making their circles of production and consumption smaller and more overlapping is important. Consumers with a degree of autonomy, whether driven by privilege, commitment, or trendiness, all get to ask, do we purchase the carcasses of decrepit birds and eggs from warehouse-raised hens or a coop for the backyard?

My own foray into conscious eating began in college, when I moved from California to North Carolina. I worked on an organic farm, and although I'd already earned my hippie wannabe card (I spent two years interning at an herbal apothecary and could recite the seven Leave No Trace principles by heart), I double-dug beds, churned compost, weeded hoop houses, and rode in the back of a truck filled with fresh turnips. Later I transferred schools, moved to Oregon, and made friends with environmental science majors who volunteered at the community garden, bucket-flushed their toilets, folded bear fat into pie crusts, and cooked roadkill raccoons in their crockpots. They talked about dumpster diving, biked everywhere they needed to go, and set up a community-wide skill share. I created a vermiculture worm bin and kept it under my desk so I could compost in my dorm room.

People are radicalized incrementally. At first you think it's just a job, then it's just friends, then you're ordering a pound of red wrigglers online and realizing you feel best in your body when it's stretching and bending under the weight of a wheelbarrow. And that's all it takes.

After college I moved back to California, tended a garden in front of the small cottage I called home, and found ways to eat meat that didn't involve corporate agriculture. I bought roosters from people I met on Craigslist. Friends gifted me their old hens who'd stopped laying. I used blood as a substitute for eggs in baked goods and harvested banana slugs from the mossy hills behind my house, then deep-fried them for dinner. I gave up ground beef and made burgers with crickets and put mealworms on homemade pizza instead of salami because insects require fewer resources than animals to produce the same amount of calories and protein.

Once you start learning about commercial agriculture it's hard to look away, and once you begin eating cricket-chickpea burgers and chocolate cupcakes made with rooster blood, you're too far gone to have any desire to go back. Alternative eating made prepackaged chicken and cucumbers that didn't come from my garden seem comparatively miserable. I began learning through labor how to belong in my kitchen and navigate my hunger, and by extension, how to exist more responsibly in my world and more comfortably in my body.

Since then I've moved across the country again, this time to southern Louisiana, and I want to do something I've never done before: raise my own chickens for eggs and meat. My cottage in the redwoods didn't have room for a coop in the backyard, but now I have the space and ability to join the ever-growing number of chicken people in the United States. Raising chickens is a natural extension of my desire to find the ways I fit into the fabric of my environment and have a life that is my own rather than something sustained by unintelligible, money-hungry forces.

From the heat-lamp-warmed brooder—a small enclosure designed to keep chicks safe and warm—four sets of wet-black eyes watch me, my own personal ultra-local iteration of backyard-to-table. The four chicks are recent additions to our household, which is a ten-minute walk from downtown and, at least on the weekends, within earshot of a heavy metal bar. It's already home to two dogs, my exceedingly patient roommate, Katie, and me—a millennial chicken wrangler trying to make up for improvisation with optimism, with admittedly mixed results. Katie is one of my best friends and closest confidants. She's watched me slaughter chickens and listened to me explain my strange dietary preferences. Katie is more steadfast, better at being straightforward, and less enthusiastic about weird things than I am, which are some of the reasons I think our friendship works so well. Her patience with me and my antics is, honestly, nothing less than saintly.

I sigh and realize the shower mat is about as good as it is going to get. I douse it with water and sponge up the excess. Staring at the chicks, I experience a bizarre mix of emotions: I want them to love me and to be useful to me. Everyone has a name. Amelia is the white leghorn with an incomprehensible penchant for flight, and Grendel is the Welsummer, a brown chicken breed from Holland, who plows past the others with plucky vigor. I name the brown Easter egger Joan because she seems whimsical like Joan Baez and start calling the blue Andalusian chick Francis because it just seems to fit her. These chickens will be laying hens, not meat birds. I do not plan to slaughter Amelia, Grendel, Joan, or Francis; I want to raise them for eggs. In a few months I plan to get a batch of Cornish crosses, who are specifically bred for their ability to convert calories to body mass, and thus meat.

In the days before the chicks arrived, I'd been hounding the local post office. I called multiple times to prepare them for the oncoming shipment of live birds, and even showed up in person to deliver my telephone number so they could call me in case of emergency, although I still cannot envision what that scenario might have been with any level of specificity. On the day of their arrival, tipped off by the company that sent them, I waited on my front steps for the mailwoman even though it was raining and ran to the curb when I saw her van. "I came here first," she told me. "They were so loud. Not my normal route."

CHAPTER 1

I brought the cardboard box inside and whisper-screamed for Katie. We crowded into the bathroom for the unboxing beside the sink. After snipping some plastic restraints around the cardboard, I lifted the lid carefully. Inside, there were four chicks, no more than a few days old, peeping on top of a bed of straw. Perfect puffballs with toothpick legs, they were shipped from the hatchery at just a day old and kept warm by a heating pack below the straw. I turned the heat lamp on in the brooder before they arrived so the pine shavings inside were already warm and ready to receive them.

One of the first things I did was to grasp their small bodies one at a time and tip their beaks into the water dish to initiate their drinking response. According to Dr. Jacquie Jacob of the University of Kentucky, researchers have observed that chicks who haven't been shown how to drink won't, even if they're thirsty and literally standing in water. I chirped and tapped my finger into the food dish. I spent the first week wondering how they would survive in the wild and tried to cut them some slack, but now that they're three weeks old, I know better. They would die almost instantly.

Negotiating around the wet spot on the bath mat, I change their water and pour more feed into their bowl. It's a careful mix of protein and nutrients, and they shuffle forward to receive it. Since the chicks arrived, I've been waking up in the middle of the night to check on them, and I'm tired. They scuttle into a ring around the feeder and peck deliberately at their food. They usually follow a pattern of eating, drinking, and sleeping for most of the day, punctuated with some stretch of mischief. As I hang my hand over the side of the brooder, one tries to eat a freckle.

The origin story of domestic chickens has been long debated. In 1868 Charles Darwin claimed that chickens descended from red junglefowl around 4000 BCE in the Indus Valley in South Asia. This opinion was tenacious, but inaccurate. Rather than originating from a singular species, time, and place, chickens drew their genetics from a combination of red and gray junglefowl and were domesticated throughout Asia at different periods. While chicken bones have been found at various archaeological sites through South and Southeast Asia and Southwest China, the oldest chicken bones ever uncovered hail from northern China.

It also seems quite likely that the egg came first. Writing for *Time* magazine, Merrill Fabry said as much: "At some point, some almost-chicken creature produced an egg containing a bird whose genetic makeup, due to some small mutation, was fully chicken." Although the question seems silly and the answer bizarre, that's how changes in species occur. The egg that became a chicken, most likely from two non-chicken organisms, was, as Bill Nye stated on Twitter, the "result of proto-chicken & proto-rooster hookin' up. . . . It's evolution."

For thousands of years chickens were used symbolically and socially: they were employed in rituals, buried whole, and baited against one another in cockfights but neither stewed nor grilled (or if they were, it was relatively rare). From Southeast Asia and China, chickens were brought westward, where they spread into West Asia, the Mediterranean, and Europe.

The Lex Fannia Sumptuaria of 161 BCE, one of the sumptuary Roman laws (laws passed to regulate consumption and luxury), banned the consumption of more than one chicken per meal and prohibited them from being artificially fattened, presumably for the sake of preventing unnecessary extravagances, although the laws were far more symbolic than actual. Roman priests used chickens to divine the future: If a chicken voraciously ate grain and scattered it with its feet, the omen was good. If the chicken didn't eat, the omen was bad (it should be mentioned that chickens were often caged without food before the ritual, so the methodology was incredibly suspect no matter how superstitious you might be).

The chicken's turn toward food appears to have taken place in the biblical city of Maresha, Israel. Around 400 BCE, it seems people began eating chicken en masse for the first time. While chicken remains in other sites were often found intact, the bones of chickens in Maresha were scraped and rutted with knife marks from butchering. This moment over 2,400 years ago was a fundamental shift in the relationship between humans and chickens; today, chicken is the most-consumed protein in the United States. The public image of chickens has, arguably irreparably, shifted from prophetic to palatable.

Americans used to be comfortable with the idea of raising their own chickens for eggs and meat. It wasn't uncommon to keep a small flock in your backyard for eggs and slaughter them for dinner when their productive abilities decreased. Throughout the nineteenth and early twentieth centuries in the United States, households, not companies or farms, produced the majority of chicken meat. People with average means and space in their backyards—not employees earning paychecks and reporting to managers—did all the raising, handling, feeding, egg-collecting, caring for, killing, plucking, and eviscerating and created a direct connection between animals and food. The idea that slaughter isn't something everyone at a chicken-laden table should know how to do is a somewhat modern concept, spurred on by industrialization and capitalism in the late nineteenth and early twentieth centuries.

As societies became more industrialized, beliefs about how humans should interact with animals shifted. Laws changed, animal production was sequestered away from urban housing, and cultural expectations concerning metropolitan life were altered.

Spaces became increasingly divided between life and work, implicitly characterizing things like physical labor and animal husbandry as emblematic of rural environments and antithetical to urban ones. The terms *production* and *consumption* became similarly oppositional, and the pervasive idea that metropolitan areas were civilized centers and animals degraded the quality of life this image promised spread like blood in water. Animals were seen as dirty and indicative of immigrant neighborhoods or lower-class housing: it must be need, the slanderous logic went, that would drive a worker in the postindustrial world to raise their dinner rather than buy it from a store. In the 1920s the US Supreme Court decided to uphold the legality of municipal powers to regulate land use, which allowed local control of chicken husbandry practices. In the period of urban sprawl following World War II, restrictions further cemented the idea that urban and rural spaces should be kept separate. Today, regulations still limit chicken ownership for many Americans, especially those in urban areas or who live on smaller lots.

As food became more industrialized, prices dropped. This trend had to do with supply, for the boilers that were produced in factories and commercial farms during the second half of the twentieth century were far more tender and palatable than the backyard hens of yesteryear with their thin bodies and tough, developed muscles, but this shift also had to do with convenience. Young, rapidly grown chickens take less time to cook, and the availability of individually wrapped legs, breasts, thighs, or wings was increasing since butchers weren't purchasing and then carving whole birds. The idea of eating meat offends some, but not nearly as many as those who are offended by the idea of killing a live animal. It seems like people at some point stopped slaughtering, and then forgot how to slaughter, their own meat.

CHAPTER 1

Traces of this forgetting can exist in pockets of the backyard chicken movement, even as the practice of raising chickens draws people closer to food production. Leslie Citroen, a businesswoman and chicken owner living in California, turned her backyard hobby into a full-fledged business that sells fancy and heritage breed chicks. She works with upscale wineries, charges thousands of dollars for custom coops, and stocks accessories like egg stamps, chicken harnesses and tutus, and hand-painted custom coop signs. She believes that backyard chickens are on the rise because people are becoming more informed about where their food comes from, but there's an invisible line that most are unwilling to cross.

One afternoon when I visited her sunny, chicken-filled backyard, she told me that people often think "it's better to buy it [chicken meat] in a package in the supermarket, while it's a shameful thing to kill your pet." Although chicken meat is consumed on a wide scale and urban chickens are growing in popularity, most backyard chicken flocks in the United States are kept for eggs only. People want to grow their own vegetables, get their own eggs, make things that are organic and wholesome and from scratch, but only if it doesn't involve bloodshed. According to a 2014 study by researchers at University of California, Davis, chickens are more likely to be "pets," "gardening partners," and "therapy tools" than they are to end up on the table. Eggs are first and foremost in chicken owners' minds, and they generally leave the messier side of chicken consumption to industries.

While twenty-first-century chicken keepers don't generally raise their birds for meat, they do incorporate some traditional values like frugality, self-sufficiency, and hard work. Dianne, who raised chickens for almost twenty-four years in Oregon, says that her chickens provide her with the "best" free fertilizer for her "massive" garden. Angela, who started her own small-scale mealworm farm for her chickens, mixes and ferments her own chicken feed. "I'm a very do-it-myself sort of person," she told me. "I believe in real food." Ryan, who spent parts of his childhood in rural South Dakota, brought chickens into his California home after his family began gardening. "We grow lots of vegetables in the summer, and with the help of our extra freezer, they last well into winter," he said when we talked over coffee one morning, months after I met him on a Silicon Valley coop tour. "We do a winter garden too, and chickens just kind of seemed to be a natural extension of our backyard farm." People like Ryan prove that food production can be enmeshed into the space and daily lives of people who are farmers not by trade but by hobby.

Neighborhoods and cities have different and individualized livestock laws, most of which seek to limit the number or feasibility of backyard chickens. A 2012 survey of municipal backyard poultry laws by Jamie Bouvier, assistant professor of lawyering skills at Case Western Reserve University, revealed that out of the hundred most populous cities in America, seventy-one regulate chicken ownership with animal control ordinances while fourteen use zoning codes. Only three of the cities in the study banned chickens outright, but most cities—fifty-six out of one hundred—have some kind of regulation that dictates how far chickens must be kept from neighboring residences. Thirteen cities incorporated setbacks of one hundred feet, which effectively ban chickens for most residents, and seven cities incorporated setbacks of over a hundred feet. A 2015 study by researchers at Boston University and George Washington University revealed that out of the 150 most populated cities in America, 93 percent allowed chickens in some capacity.

Regardless, people like Ryan can work around regulations. Backyard chickens are still on the rise, partially because the style of living they exemplify rebels against modern metropolis ails. In the wake

CHAPTER 1

of stresses about increasing urbanization, environmental collapse, GMO foods, and kids growing up with their fingers on screens instead of in the dirt, chickens are an all-inclusive reprieve. Chicken people tend to have concerns about the environment, industrial food, and the economy of commercial agriculture. By owning chickens, they perform a feat of micro-resistance against society's dominant forms of consumption and production and create a counter-narrative to the story that food, something we all require on a daily basis, can only be produced by certain industries in sequestered places.

Projects like canning vegetables, raising chickens to avoid buying eggs from a store, or planting apple trees along property fences challenge the idea of what is appropriate and inappropriate for suburban life. The ideologies brought about by industrialization fundamentally changed the way Americans engaged with food production and, in part, led to the chicken's privileged position as the food source it is today. From 1909 to 1940, when most chicken meat was coming from backyards, the per capita consumption of chicken for Americans remained around ten pounds annually, fluctuating by a pound or so, but a shift occurred midcentury.

It would be impossible to tell the story of chickens as an American staple without mentioning the entrepreneurial skills of a twentieth-century woman in Delaware. From pictures she looks pragmatic: short hair, glasses, round face, broad nose. Her image rings farmwife, not foremother of a multi-billion-dollar industry, but that she is. In 1923 when Cecile Long Steele ordered her annual fifty chicks from Dagsboro Hatchery, they sent her five hundred—a clerical mistake she didn't dispute. Cecile normally sold eggs to supplement her husband's income from the US Coast Guard in Bethany, Pennsylvania, and feeding that many chicks into adulthood would've been too expensive. Cecile raised the chicks until they were two pounds each and sold them to her neighbors for meat at 62 cents per pound, for a profit of around $775 (equivalent to about $11,450 in 2019). Her husband quit his job to work on his wife's pop-up chicken farm, and three years after the legendary five hundred chicks, Cecile had a successful startup that involved a henhouse that could accommodate ten thousand birds. We'll revisit Cecile again later, but she is worth mentioning here as well; her legendary chicken farm set a precedent that has reverberated through the decades. It's quite possible that if Cecile hadn't begun selling young, tender chickens for meat someone else would have, but in fact that isn't how it went.

Prior to Cecile Long Steele, chicken meat was generally the result of old hens and unwanted roosters; it was not something you'd buy. But other farms caught on, the commercial industry blossomed, and in 1952 broilers (chickens who are selectively bred to grow faster and larger than their egg-laying counterparts) surpassed backyard-raised chickens as the main source of chicken meat in America. A few years later, in 1958, annual per capita chicken consumption jumped almost 2 pounds from the previous year, the largest uptick recorded since 1909, and continued to rise. In 2018 the average American consumer ate a whopping 93.5 pounds of chicken.

Whether from sheer volume or from environmental effects, it's become clear that the predominant ways of producing food are no longer sustainable. The standard response from corporate agribusiness would ask us to consider how the carrying capacity of the planet has increased, and how today more people are fed, and fed better, than they were two hundred years ago. We'd be asked to examine how the industrialization, specialization, and mechanization of agriculture have allowed for greater yields with fewer inputs. The diversified farms that produced a variety of vegetables and livestock and

CHAPTER 1

later transformed into sweeping monocultures could be considered a feat of corporate agriculture success, but the world is paying the price for cheap food. Look no further than the ways agriculture has decimated the United States' topsoil, which, according to a 2006 article in *Environment, Development and Sustainability*, is lost at a rate ten to forty times faster than it can be naturally replenished and comes with a $37.6 billion annual price tag by way of "productivity losses," a fancy way of saying we're diminishing the quality, capacity, and fertility of the land. Look no further than the low life expectancy and high infant mortality rates in North Carolina communities near commercial hog farms, the antibiotic-resistant strains of *E. coli* found in chicken meat worldwide, or the video footage of Tyson employees throwing and hitting chickens with their hands and impaling chicks with a nail on the end of a pipe. Look no further than the workers who are maimed and killed in unsafe slaughterhouses or the unreasonably high rates of employee turnover, which sometimes exceeds 100 percent annually. A 2015 report by Oxfam revealed that line workers at Tyson Foods, Perdue Farms, Pilgrim's Pride, and Sanderson Farms poultry production facilities were denied breaks during long shifts, so some resorted to wearing diapers. Ray Offenheiser, Oxfam's president, stated in a press release that "poultry workers are among the most vulnerable and exploited workers in the United States." These industries leverage the US immigration system by hiring undocumented workers at disproportionately high rates on the assurance that this will allow mistreatment and exploitation to continue.

In 2017 the *New Yorker* published "Exploitation and Abuse at the Chicken Plant," which reported on the injustices by Case Farms. In 2015 at a poultry facility in Canton, Ohio, a worker named Osiel López Pérez, who was only seventeen at the time and too young to legally work in a factory, was injured while cleaning factory equipment. After the incident, which resulted in a lower leg amputation, several underage and undocumented individuals, including Pérez, were fired. According to the article, "Case Farms has built its business by recruiting some of the world's most vulnerable immigrants, who endure harsh and at times illegal conditions that few Americans would put up with. When these workers have fought for higher pay and better conditions, the company has used their immigration status to . . . quash dissent." In the same year that Osiel López Pérez was maimed, Case Farms' plant and hatchery in Goldsboro, North Carolina, were given an Award of Distinction from the Poultry Industry Safety and Health Council.

It isn't just Case Farms.

In 2019, the *Washington Post* reported on a rash of ICE raids at a number of poultry plants in Mississippi, which led to seven hundred people being detained and constituted what was described as "the largest single-state immigration enforcement action in U.S. history." This was not by chance: research shows that, beginning in the 1990s, the poultry industry began soliciting Latin American immigrants to come to Mississippi. According to the National Center for Farmworker Health, approximately 50 percent of workers involved in poultry processing are Latino, and 25 percent aren't legally allowed to work in the United States.

Our food system is sick, and the environmental and social ramifications are staggering. New styles of sustainable agriculture often focus on smaller, more individualized operations than the massive corporate systems they seek to replace. Systems of alternative production and consumption—low-impact farms aware of their carbon footprint, rogue squat gardens nestled illegally into the fabric of

CHAPTER 1

cities, cricket farms in warehouses, and restaurants that serve nose-to-tail meals—are urging people to radically recalibrate their understanding of food and land. As Wendell Berry observed, "How we eat determines, to a considerable extent, how the world is used," and we are reimagining eating.

Will we glut ourselves on cheap crops and let the world go the way of our depleted topsoil, or will we invest in something long term? What we buy determines what and who hold power over us. Sometimes people cannot control where this power goes because of financial and systemic pressures, feeding into systems that undernourish individuals and communities. Under the strain of limited finances or a lack of access to grocery stores and fresh food, the options that individuals have can be limited and their choices functionally nonexistent. McDonald's hamburgers are cheaper than organic carrots. Limits like these can result in a Coke or Pepsi dilemma: the illusion of choice between carefully curated options that essentially mean the same thing. But other times, our autonomy as consumers isn't negated by systemic factors, and we do have the ability to decide what will hold power over us. The level of privilege that allows people to choose things like going to a grocery store versus a convenience store, buying organic instead of conventional, and raising chickens in their backyard cannot be understated.

Again it's 6:00 a.m., and again I haven't had coffee, but I'm kneeling beside the warm brooder. I'm swabbing the chicks' butts with a dampened Q-tip because they are prone to a condition called pasty butt, which is exactly what it sounds like. It can be fatal within a few days if left unchecked, so it's important to closely monitor their rear vents.

My dogs, Tashi and Atlas, whine from outside the closed door because I won't let them in and haven't fed them breakfast yet. I remind them through the wood that I haven't eaten either and grab Grendel, who peeps in sheer desperation and kicks her small toothpick legs. I wipe a clot of dried feces from her vent before placing her back on the pine shavings. She runs to the other corner of the brooder.

"I am so sorry," I say to Amelia before snatching her. "Please forgive me." She screams with unbridled terror. I dampen the Q-tip again and clean her vigorously, gently place her back with the others, and seize another chick from the brooder. "Please don't remember this," I beg. "Please love me still."

During the domestication process the bodies and brains of wild animals undergo a series of radical changes. Some of the shifts modern chickens' ancestors experienced were small and relatively insignificant, like the development of yellow skin, but others were substantial. Domestication often increases an animal's response threshold, or the point at which they will react to stimuli, which makes things like handling them and putting them into relatively close quarters with their kin less stressful on the creatures and easier on the human overseers. Although these chicks have been shipped through the mail, manhandled, plunged beak-first into a water dish, and repeatedly shuffled to boxes as I change the pine shavings in their brooder, they seem relatively unfazed. It is altogether possible that this preventive treatment will pass like their other fear-based memories.

In a comparative study published in *PLoS Genetics* between the stress responses of white leghorns (one of the most popular egg-laying hens today) and red junglefowl (the wild genetic kin of today's modern chickens, whose DNA more closely resembles an original

CHAPTER 1

shared ancestor), researchers noticed fundamental differences in the brain functions of the domestic verses the wild birds. After exposure to stress, the red junglefowl had higher levels of corticosterone and corticotropin-releasing hormone (CRH) than the leghorns did. CRH is heavily involved with the creation and storage of fear-based memories, suggesting that modern domestic chickens don't become too dismayed when exposed to previously traumatic stimuli, or at least, they don't become as dismayed as the red junglefowl, and therefore they're perhaps less sensitive than their genetic predecessors. While this fear-obliterated bliss isn't total—chickens can and do get scared, and you can lose their trust if they perceive you as a threat—they're at least less skittish and more likely to recover from distressing events than red junglefowl.

Domestic chickens are vulnerable, needy creatures, their ability for adaptive learning in the wake of fear-based memories aside. The chicks live in the bathroom and take over. What first appeared to be four unimposing, fragile creatures prove to be four raptor offspring. I give them a dust box, which they immediately take to. Dust bathing is one of the behaviors that makes a chicken a chicken: it's instinctual. In older hens, this serves the purpose of preventing and killing external parasites and removing excess oil from their feathers and skin, but the chicks are too young for these problems. Instead, they appear to just do it to have fun. They ravage the small cardboard dust bathing box. Their small claws rip out the bottom, and although they enjoy themselves endlessly, a layer of dust begins to coat everything in the bathroom.

When they aren't making a mess, they're finding ways to get free. I constantly add height to the brooder by attaching new sections of cardboard every time I find Amelia—it's always Amelia—chirping fiercely as she stands on the blue shag bath mat.

Every week I raise the heat lamp, lowering the temperature in their enclosure a measured five degrees, and every week I make the brooder larger, adding various wings. I move them from one corner of the bathroom to the other. They block the bath tissue and then the sink. To get to the toilet I have to shimmy around them, and when I want to brush my teeth I have to lean far forward to get to the sink, making my body a top-heavy L. Exiting the shower becomes a precise practice in avoidance, lest I step into the brooder or get the cardboard wet.

For those who have no inclination to fill their bathroom and then backyard with livestock or turn their lawn into a vegetable forest, it might seem hard to imagine why someone would willingly sign up for this kind of work. During a conversation with my friend Geran Wales, who worked as a member of a permaculture collective in Oregon that created a community garden on an abandoned lot, he told me there's a reason the term *animal husbandry* exists. "At one point, people knew that you had to take care of these animals if you wanted the benefits that living with them provides."

Changing the chicks' food and water, swabbing their butts, crawling into the bathroom one, two, or sometimes three times a night because my dreams are suddenly ravaged with images of falling heat lamps and collapsing walls: this is my version of animal husbandry. I find myself increasingly bound to the chickens and to this style of care. When I'm tired of the mundane repetition of cleaning up after them, I envision real mayonnaise, something I haven't consumed in years after I stopped eating animal products that came from commercial agriculture.

It seems natural that the more something means to you, the more effort you'll put into getting it. It feels like I've been up to my elbows in chicken care since they arrived, although I know this will only get

CHAPTER 1

worse. *Mayonnaise*, I tell myself, *I'm willing to put in time and labor for mayonnaise*. This, the inescapable clencher to involving myself with the production of what I consume and the small truth at the heart of raising animals in this capacity. If I want it, I work for it.

What is the alternative? It seems that animal husbandry can be a balm to the current state of the poultry industry. According to Bernard Rollin, an American philosopher and professor who helped draft the 1985 amendments to the Animal Welfare Act, "The way in which egg-laying hens are kept in battery cages is arguably one of the worst, if not the worst, of many inhumane methods in modern agriculture." While it's easy to imagine that the organic eggs you buy from the store come from pasture-raised happy hens who spend their lives pecking in the dirt and kicking up bugs, this image isn't reflective of reality. "Traditionally," states Rollin, "95 percent of egg-laying hens were maintained in small wire cages, with as many as six to a cage. Each animal therefore had less space . . . than that occupied by a sheet of typing paper, and in common situations which I have observed, chickens lived on top of other chickens."

When I visit Isabelle Cnudde, she places a one-eyed chicken named Curry in my lap. A software engineer turned chicken rescuer and animal advocate, Isabelle has dark hair, brown eyes, and delicate features. She doesn't wear any makeup and has a direct, practical air about her, but her kindness is clear. As Curry's body heat warms my legs, Isabelle tells me that a third of the hens she rescued from an organic, free-range farm were missing eyes. Curry was one of those hens, and her remaining pupil is shrunken and malformed. "People think, 'Yeah, it's great, because they don't put chemicals in their food and whatever,' but then the chickens don't have medicine," Isabelle says. "Infections happen, they don't get medicine or anything, and then lose the eye."

It's easy to ignore the reality of something as benign as eggs. After all, people think chickens don't die for eggs—they're just something hens pop out on their own, but that's not the case. The reality of egg-laying chickens is often a bleak one. Chickens are killed prematurely when their egg-laying drops, often debeaked to prevent outbreaks of cannibalism, and kept in confined, unsanitary conditions. Even if egg laying is a natural process for chickens, store-bought eggs signify the lives of animals who lived under duress and, best case, were gassed on-site at around eighteen months old.

When you look beyond the egg to the animal that produced it, as backyard chicken owners do on a daily basis, purchasing decisions suddenly seem more worthwhile. Knowing what goes into your food dissolves myths created to sell more eggs and make consumers comfortable with products and practices that would otherwise trouble them. Understanding the methods of production allows consumers the possibility of actual choice. People can decide whether or not the ends should be separate from the means: Are gassed chickens worth their eggs? Are pasture-raised eggs worth the price tag? Some say yes and some no, but without a realistic understanding of production inputs and practices, consumers aren't given the capacity to decide for themselves.

Know what goes into your food; otherwise, there's no possibility for real choice.

When the chicks are six weeks old, they're ready for the outside world. I move them to the coop and construct a large pen in our backyard with metal stakes and bamboo fencing. They are all fully feathered. I've been taking them outside every morning so they can

CHAPTER 1

peck in the grass and scratch for bugs. With a blanket, book, and cup of strong coffee, I supervise in the form of taking pictures, digging up earthworms, and not reading.

"Be safe," I tell them before latching the coop door the first night. "Sleep well." It's late September and the air is thick with humidity and mosquitos. They've spent the entire day outside, the brooder is finally dismantled, the bathroom is cleaned, and the chicks have extra water stashed on the second floor of the coop. After closing both latches, Katie and I sit on the back patio and have a beer. We've decorated the covered concrete slab with string lights and potted plants in addition to patio furniture inherited from the tenants before us. While we sit on the chairs under the electric glow, I can hear peeping. I interpret this as distress, although Katie assures me that they're probably fine in the coop. Katie is more practical than I am and a bit more normal. She doesn't do things like scan Craigslist for unwanted roosters or old hens. She's a recycler extraordinaire, has raised bees, and supports my strange chicken antics, but she doesn't share her bathroom with animals. Listening to the chicks, I realize all at once that they have never been in complete darkness.

We're still outside after a second beer, and I'm still listening. Night has fallen, creaking with the sound of insects, and the backyard is dark despite the full moon due in five days' time. I pull on a headlamp, tiptoe to the coop, and peek through the wire doors. The chickens are huddled on their perch, swarmed by mosquitoes. They blink into the blinding light.

I go inside and do not sleep well. I wake up early the next day and immediately head outside in my pajamas, squinting sleep out of my eyes, to let them out of the coop. I feed them, change their water, and then return shortly after with the pureed dregs left over from making oat milk.

I coax them over with the oatmeal mush, let them eat from my hand, and then pour water into their chicken feed since they seem to like it better wet. Grendel investigates a fleck of dirt on my hands, Amelia stretches her wings and charges across the run, and Francis and Joan begin to dig and peck the grass. I put my coffee on a stump, grab the shovel leaning against the fence, and dig until I find june bug larvae.

This is the aftertaste I want my future breakfasts to leave behind: frantic wings, curious beaks, dew, dirt, worms, sun. Living things being living things.

2

Biking for the Birds

SILICON VALLEY'S TOUR DE COOP

A FEW WEEKS LATER on a Saturday in September, I'm in Silicon Valley, California, with a backpack full of water, snacks, and cash, riding my mom's old Schwinn cruiser in the wrong direction despite the route taped to the handlebars. I'm in my third year of a graduate fellowship at a large university, which means I'm done with coursework and coasting along in a gap year between teaching requirements. My flexibility is unique, allowing me to drive to Houston, catch a cheap flight to San Francisco, stay with my parents, and attend Silicon Valley's Tour de Coop in the name of research. That morning I loaded Mom's bike into a borrowed car, drove to Palo Alto, parked in one of the recommended locations, and set off—trying to honor the spirit of the bike tour as much as possible without riding the extra sixty-five miles it would have taken to get to Silicon Valley.

The chicks not only survived their first night but they've adapted to life outside. In the mornings I use a shovel to dislodge worms from the dirt, and by the afternoon they usually take to dust bathing in the dug-out holes. When it starts to grow dark they get themselves back into the coop, where they will roost on their perch until morning. There have only been a few nights that I've "lost" Amelia only to find her roosting in the fig tree.

In my short absence, Katie agreed to watch not only the dogs but also the chicks. The tasks of dog and chicken maintenance are mostly similar, involving feeding and watering the creatures and dodging them as they run around your feet for whatever reason. The dogs have already made it clear that they prefer Katie's bed to mine, and she's always gracious enough to accommodate them. I arrived in California with confidence that the animals were in good hands.

Silicon Valley, located at the southern end of the San Francisco Bay Area, is one of the state's most affluent communities. It's a global metropolis of cutting-edge technology, home to the headquarters of Google, Apple, and Facebook, in addition to a plethora of start-ups, young tech millionaires, and most recently, backyard chickens. Today's bike tour is an annual, free, self-guided trek to see backyard chickens, fancy coops, homesteads, beehives, and gardens.

I double-check the directions, make a U-turn, and start heading uphill. Two families biking together haul toddlers in front-facing handlebar seats and young kids in blue-painted trailers. They outpace me; the children wave at my receding form. The kids and adults have chicken decals taped to their helmets; the white chickens with yellow feet and red waddles look like dorsal fins, and the group parts the wind like a school of crafty, bi-species land sharks. I missed the DIY memo. I'm sweating and only momentarily, laughably, try to keep up. The adults pedaling have quads like steel cables.

CHAPTER 2

This year, nearly 1,800 people registered to participate. People in the community volunteer to host coop stops where they open up their backyards, put up lemonade stands, answer questions about coop construction, and show off their urban goat barns. There are twenty-nine coops on the tour, a small fraction of what Silicon Valley has to offer.

I pass wide white gates standing sentinel before large mansions. There are front yards with fountains, manicured lawns, and landscaped swatches leading up to garages the size of large homes. There are rows of lavender and trellises of pink roses. I wheeze and—out of a combination of frustration and sensitivity—imagine the children's faces bobbing before me, expressing seeming perplexity as to why I was so slow. Somehow, chickens are not out of place in this neighborhood, but on my mother's old Schwinn with the broken gear shift and unusable back brake, I feel like I might be. I stand up on the pedals to exert force on the tires as an incline rises under me.

I'm not sure I understand why people making a ton of money on their computers all day want to come home, pop a microbrew, and enjoy the dulcet sounds of satisfied clucking. From my experience, coming home to chickens means throwing my bag on the floor, trading dress shoes for galoshes, and cleaning the coop, watering dish, and feeder. It's romantic, sure, but it's also laborious. Some people understand this phenomenon a little more than I do. Still standing on the pedals, passing multi-million-dollar mansions, I think back to my meeting yesterday with Leslie Citroen, who started her own thriving chicken business. Leslie sells chickens and chicken feed, designs custom coops, and works as a "chicken consultant" (yes, that *is* a thing, and the going rate for a chicken whisperer of Leslie's expertise is $225 an hour). She lives in Mill Valley, a town just north of San Francisco, but a lot of her clients live in Silicon Valley and the South Bay.

In some ways, Mill Valley is a lot like Silicon Valley itself, minus the multitude of headquarters and emphasis on high tech. From 2000 to 2016, the median household income increased from $90,794 to $154,599, with approximately a third of the households making over $200,000 annually. As incomes increased, the number of backyard chickens seemed to rise as well. Leslie sells more rare and exotic heritage breeds today than she did ten years ago, and her prices—while not necessarily high for the chickens she's providing—do reflect the affluence of the areas she serves.

Leslie charges up to $52 for a chick. Depending on the breed, a six- to eighteen-month-old chicken could cost between $65 and $189.99. People are happy to pay those prices. In some ways, she's the only game in town if you want fluffy-footed, blue-egg-laying designer chickens. It's as simple as supply and demand. "I stopped trying to compete with the feed store," she said. "They just have the traditional Rhode Island reds and barred rocks, so I work with many more breeds. That's how I specialize, and that worked for me."

While Leslie has devoted most of her yard to her flock of twenty to twenty-five free-ranging hens, she is also a realistic businesswoman. As we sat on her back porch drinking iced tea under the shade of redwoods and fruit trees, she told me, "It drives me batty that most of the people with chickens are 'the Chicken Lady.'" The misnomer is applied to her often, and given her expertise, backyard full of well-behaved exotic and heritage breed hens, and public image, it's possible to see why. When she clapped, her chickens came running in anticipation of treats and grains. But Leslie isn't the feather-loving equivalent of a cat lady; she's a serious businesswoman who's made a living out of the backyard chicken market at an opportune time when few were realizing the impact chickens would have. Leslie is slender and has brown hair and tan, strong-looking arms. "Chicken broker," she corrected with a smile.

As Leslie understands it, there are a few reasons the popularity of backyard chickens has surged in places like Silicon Valley and the Bay Area. For one, they aren't very difficult to take care of and essentially have the independence of outdoor cats. They give you something substantial: provide them with food, shelter, and water, and you'll get eggs. In Northern California the climate is particularly conducive, and with only a few nights of frost a year, chicken owners don't need to worry about keeping their hens warm during the winter. Leslie told me that "people in this area are extremely well educated and very knowledgeable." They won't balk at expensive coops or organic feed, and in addition to having money to spend on their chickens when the flock is healthy and thriving, folks can shell out thousands of dollars on medical bills if something goes wrong. "In other parts of the country, to spend that much money on chicken surgery is beyond the pale."

While you don't need to be affluent to raise backyard chickens, it can help. The Bay Area is a world of Whole Foods, small batch kombucha, and fixed gear bikes worth thousands of dollars. It isn't hard to see how money can support liberal beliefs about politics and the environment. Leslie believes that a mild climate, education, and money make a big difference when it comes to raising backyard chickens. I bought Amelia, a white leghorn, from a hatchery for about three dollars, but that's not quite the Bay Area mentality. Disposable income equates to flexibility. People might use this openness to live in alignment with their environmental and personal values, but in the land of biohacking, super competitive startup ventures, and buzzwords like *optimization*, a chicken isn't just a chicken.

"Everyone knows factory farming is bad," Leslie said, "but very few people turn vegan or stop. People are very wealthy here, and they are willing to spend more money for their food to make themselves feel less guilty." Buying organic chicken feed, commissioning someone to build a custom coop worth thousands of dollars, and having a flock of fancy heritage-breed chickens in your backyard can be a guilt-free alternative to the commercial egg industry. Leslie made it very clear that backyard chicken owners in the Bay Area care about their chickens and treat them responsibly. Although raising backyard chickens for eggs can be a type of alternative eating that sidesteps some of the more harmful impacts of modern agriculture without giving up too much at the table, this isn't to say that people don't love their chickens or that human owners treat their chickens as a means to an end. On the contrary: the chickens I meet during the Silicon Valley Tour de Coop seem to be living more like pets than livestock.

I have completely lost track of the crowd of fellow Tour de Coop bicyclists and am terrified of cars on the road. I take a winding, steep grade with little-to-no shoulder through a wooded neighborhood. My mother's bike is unwieldy, and I hope the flat tire I filled the night before remains firm. After cresting one of the many hills, I take a sharp left into a driveway and end up at a house with a five-chicken coop in the backyard. The house is stone and looks like something out of the French countryside.

After leaning the bike against the bushes, I hang my helmet on the handlebars and run my fingers through my sweaty hair. Under the cover of the wraparound porch shaded with ivy, I notice a

CHAPTER 2

table with a spray bottle of Lysol and directions to please spray the soles of my shoes before going any farther to minimize the risk of cross-contamination with another nearby coop. First things first: biosecurity.

The path to the backyard leads through the covered walkway, up a staircase, and past a stone fountain. There is a long table on one of the porch terraces where two kids—presumably they live here too—sell cookies and hand out cups of water flavored with mint and cucumber.

The coop is custom built specifically for the yard, with rat- and predator-proof concrete sides, drapes near the nest boxes that match the wreath on the door, hanging treat-dispensers, and specially engineered slanted nest box floors that allow eggs to gently roll out from under the chickens, down a chute, and into a shelf along the back of the coop where they land upon stretched canvas for easy collection. The woman who lives here tells me she bought the chickens and coop as a sixth birthday present for her son, who opted for chickens over a trip to Legoland.

"He's very shy and uncomfortable talking to people," the woman says, "and so for him I think this is great because it's something he's comfortable talking about and because it's his." This is clearly the case. As he passes cookies for cash across the covered table, he talks about the hens that laid the eggs. When I buy a cookie and get a cup of water, he mentions that his chickens love mint, which is why he grows it in the garden.

For his mother, the chickens give her home a "whimsical quality." She loves having so much life around her. The tapered stone steps that lead to the coop, the archways dripping with overgrown foliage, and the coop itself, decorated with fabric wreaths and twinkle lights, are undeniably fairytale. The benefit of chickens goes beyond fresh eggs and extends to teaching her son about how to treat animals, encouraging his growing confidence, and making her home feel like an oasis. The space looks less like a yard and more like a secret garden.

At the next house, which is mercifully less than a mile away and up and over one small hill, plus a staggering driveway I elect to walk my bike up, I meet Ryan. He's an executive at a high-tech firm and manages a global team, overseeing people in India, Costa Rica, Ireland, Atlanta, and Palo Alto. Surveying Ryan it's clear: he wouldn't have had a problem with the hills. Ryan is tall, middle-aged, and muscular. He bikes, he swims, he runs. He's done the Escape from Alcatraz Sharkfest swim three times. When he moved to Palo Alto eight years ago, he began to think about balancing his work and personal life, especially considering the global reach of the company he works for. "Earlier in my career I took calls at all hours of the day, but now, unless it's extremely urgent, I don't take calls after 11:00 p.m. or before 4:00 a.m. because you've got to get core sleep." Chickens have become part of this balance for himself and his children.

His coop has an attached shed area where feed is stored, and although his flock is sequestered into its coop for the tour, the eighteen or so chickens eye the vast field they normally free-range in, which runs along one of the lower sections of the property. There's a seed ball hanging in the spacious walk-in coop. The chickens peck it regularly between crooning and scratching the earth with their claws and seem content, for now, to stay inside. Along the back of the coop are doors to access the nesting boxes.

When I ask Ryan about his motivations for keeping backyard chickens, he says, "It's a very affluent area, and I think it's important for kids to get their hands dirty, like I did as a kid, and not just have everything handed to them or spend all of their free time on devices." His kids care for the chickens, sell eggs, and haul feed.

CHAPTER 2

When a hawk attacked one of his chickens, Ryan decided he wasn't going to seek medical care for her. "Some of our other friends with chickens only have four or five, and they're more like pets, so when they have issues they go to the vet," he says. It's not to say his neighbors aren't pragmatic, just that he never intended for his family's chickens to be pets. They're a job for his kids, manure-makers for the garden, and egg-layers for the kitchen, but they're not necessarily companions first and foremost.

"They're not going to be in the house like a cat or dog that's always protected," he tells me. "They're somewhere in between pets and livestock . . . but we love them all." Initially, the chickens weren't given names, although that's changed over time. Ryan's family doesn't eat their chickens—they're not *that* much like livestock—but they do eat chicken at home in addition to the eggs their hens produce. His kids have made the connection between animals and food through these twin tracks. "They also see the end-of-life process and know the 'pecking order' is a real thing," Ryan tells me. His children, both teenagers now, were sad when they lost their first chicken, but "just like anything you get used to it, and you expect it." Predator attacks are often part of having free-range chickens: it happens.

Despite the attacks, Ryan thinks it's worth it to let his chickens free-range. His neighbors, for the most part, don't. When I ask why his opinions differ, he tells me, "Probably more freedom." He wants his chickens to be given the right to roam. "You'll see a lot of controversy on how chickens are raised, and you see the tiny spaces they have, and our coop is plenty big. By commercial standards we could probably have fifty chickens in there or more, but we want happy chickens. Plus, by free-ranging our birds, I think we get more tasty eggs." Almost anyone who has eaten eggs from backyard chickens will tell you that backyard chickens lay better-tasting eggs, and it is absolutely true that they are more nutritious. In 2010 *Renewable Agriculture and Food Systems* published a comparative study of eggs from pasture-raised versus caged hens; the former had a 38 percent higher concentration of vitamin A, twice the amount of vitamin E, more than double the amount of omega-3 fatty acids, and a far healthier ratio of omega-3 to omega-6. Backyard chickens, who are often allowed to forage, can produce better eggs than factory-farmed hens.

While chickens' foraging abilities dwindle in the winter because the days grow shorter, colder, and wetter, during past summer months Ryan's flock has been gorging themselves on bugs and grasses.

Even if they aren't pets, "they are therapeutic just to have around," Ryan says. He talked to his neighbors before moving the chickens to their side of the property, and they said they loved hearing the clucking sounds. "Our friends, especially people coming from more urban locations, haven't always seen chickens." The naked neck, who Ryan describes as his "friendliest chicken" and compares to a golden retriever, always makes a good impression. She will walk right up to strangers. Naked neck chickens are named for their most obvious trait: a long, scrawny neck draped in what can only be described as "chicken skin" instead of feathers. Ryan's chicken's plumes reemerge atop her head, giving her the curious look of a flamingo wearing a hat. "It likes to be held," he tells me.

Later on the tour, I visit a modern homestead with goats, buy a small chunk of beeswax, and wander through a raised bed garden with a long coop where I don't dare attempt small talk with the women wearing floppy hats. They sit around a table in the shade, drinking glasses of white wine, and I can only guess that one of them owns the home, but I have no idea how to politely introduce myself. I feed an alpaca, get lost on my bike (again), and feel cheated by the incline. The houses are expansive and yards manicured.

CHAPTER 2

At the front door of one house, the homeowner, whom I will call Stan (this is not his real name), corrals a group of us into a tight pack. "I'm not letting you in until I tell you four things! I'm Stan," he says, "nice to meet you. Would you please at some point sign in? I will happily talk to you about electric cars and the house," he says. "So if you sign in, I'll do the talking. I welcome the Tour de Coop!" We bottleneck at the entrance. "I'm going to wait until those people come," he says as a couple walks up the sidewalk to his front gate, "because I'm the homeowner, I'm going to do the pitch, and my voice is going."

When it's my turn I take the clipboard and wonder, *why am I being asked for my email address?* There's a box to check off: are you a Palo Alto resident? I cross through "No" with a ballpoint pen.

We finally get to the Four Things: energy-efficient houses, electric cars, local politics, and chickens. I listen to him explain how his passive, energy-efficient, low-carbon-footprint house came to be and why it's so functional; give his pitch about why we should all consider getting electric cars and how they're actually really affordable and fun to drive; and encourage us to vote for so-and-so for city council because of such-and-such reasons. He does not actually discuss chickens and I tune out, which is ungracious and judgmental, but standing shoulder-to-shoulder on the front walkway listening to a self-satisfied sermon on sustainability in the late summer sun when I have another coop to visit, a sandwich back in the car, and no intention to vote in the Palo Alto election or buy a Tesla, I want to be a jerk.

As we walk inside, Stan describes the house's heat-recovering ventilation system, the heat exchanger, the grape arbor that grows lush in the summer to provide shade and loses its leaves in the winter. A few of us spread out, but I conform to social niceties and keep smiling. He tells those of us who've continued to listen that the sun the winter-bare arbor allows to come in through the large sliding-glass doors heats the concrete floor, which helps keep the whole house warm.

The walls inside are an inviting shade of orange, and the long, simple dining room table conveys a kind of farmhouse ease. It's a beautiful home and admittedly impressive, although I can't avoid the feeling of being talked at. There are laminated plaques on the walls describing different energy-efficient functions, and it seems to me like they are permanent installations. Stan mentions that he's had over three thousand people come through the house and that it uses about 80 percent less energy than a typical house, and I get the vague impression that this space is a museum unto itself as much as it is something to be lived in.

Houses like this—with specifically designed walls to capture and release heat and systems in place to control lighting and cooling without using any energy—might be the energy-efficient homes of the future that we need. Stan has an impressive and forward-thinking home, and regardless of his methods, he's generous for sharing it with us. As the pitch slows and the group spreads out, Stan offers us food. "If anyone wants eggs," he says, crossing into the open-floor-plan kitchen, "I'm making eggs."

When no one says anything, he continues. "Sadly," Stan says, "they're not our chicken eggs because at this point in time the ladies are a little bit old, so we're eating more eggs than we are producing." He pulls a cardboard egg carton from the fridge.

I'm struck by this blend of seriousness and playfulness. He'll make eggs for strangers, clearly cares about reducing his carbon footprint, and wants to promote environmentalism, but he keeps chickens that are not productive layers. This means that they cost him feed and space, which equates to resources in the forms of any inputs necessary for manufacturing their feed as well as occupying a coop that might be filled with more productive pets. Their carbon footprint becomes Stan's carbon footprint, plus that of the hens who laid his (presumably) store-bought eggs.

Stan appears oddly contradictory to me, like a particular brand of low-key high-key, of being so Zen you will cut someone off in a health food store parking lot to get your alkaline water before Pilates, of being such an organic locavore that you keep chickens in your backyard but pay someone else to clean the coop (Leslie Citroen, incidentally, mentioned including this service in her business when we spoke).

When I finally make it to the backyard, Stan's intern talks about the coop and shows me how easy it is to collect the feces that falls under the nighttime perches. His intern isn't a chicken intern per se, but I suppose when you've reached the echelon of life that Stan has, someone will shadow you in their free time.

The chickens are three years old, so they're past their prime production years. However, the intern tells me, the chickens are "part of the spirit of the house."

I thank her, thank Stan, and pass his Tesla on my way out the front gate. I rush to make it to Clorofil, my final stop for the day.

Clorofil is a farm animal micro-sanctuary run by Isabelle Cnudde in her backyard. Isabelle uses her platform to inform people about chicken health, chicken adoption, and plant-based lifestyles. At Clorofil, Isabelle rescues, rehabilitates, and rehomes chickens from egg farms. Laying hens are generally killed when they're eighteen months old and their egg production drops. "Egg layers here in California are gassed and trashed to landfill," she tells me.

This seems like a strange practice—throwing away chickens when they are one of the most commonly consumed animals in the United States—but laying hens don't fit with most consumers' ideas of how chicken is supposed to taste. Laying hens are much older than broilers when they are killed—a year and a half compared to thirty-five to forty-nine days—so their flavor and texture are different than the arguably juvenile birds that Americans are used to eating. As a populace, we've come to expect one thing from chicken, but that's only because we've been eating birds that aren't even two months old. There isn't a huge market for spent hens.

Although at one point spent hens went into Campbell's soup, this changed over time. According to Chad Gregory, president and CEO of United Egg Producers, "Due to losing market options, the US government purchase programs for school lunch, military, and prisons have become the largest single buyer of spent hen meat."

Spent layers are also sometimes used in pet foods. While this might be better than filling landfills with the bodies of unwanted hens, it's arguably less humane. In a 2018 article published in *Huff-Post*, CEO of Vital Farms Matt O'Hayer claimed that getting spent laying hens from a farm to a slaughterhouse involves "pack[ing] them into crates and ship[ping] then in trucks hundreds of miles. It's stressful for them and it's not very humane." Although CO_2 gas is quicker, the process is more wasteful.

It's easy to imagine that producing eggs is more humane than producing meat, but this is a work of fiction, a poorly concocted consumer mythology. If I want to eat eggs, it seems best to get them from my backyard. I want to know what goes into my food.

This isn't just about ingredients. This kind of knowing means becoming familiar with what the production of food requires. Without firsthand knowledge—watching video footage snuck out of a building owned by the company you buy those convenient cartons from or going to an egg farm—or secondhand knowledge, like investigating the meaning behind certifications like "free-range" or "cage-free,"

CHAPTER 2

reading about what chickens experience, or talking to a farmer, consumers don't actually know what goes into their food. They might know the ingredients, but they don't know what the production of their food means or requires, and that's not an accident.

Owning chickens can change someone's perspective, as it did Isabelle's. Seeing what dynamic creatures chickens can be and understanding that they have needs and express preferences can help consumers connect the dots that eggs come from chickens and chickens in most industry operations are mistreated. When she transitioned from being a software engineer, Isabelle knew she wanted a change. "I decided that, although I was making a lot of money, it was not what I wanted to do, so I retired from the corporate world and decided to . . . dedicate my life to chicken care."

Most of her backyard is devoted to an expansive chicken run with a large coop, a secluded area for a six-year-old blind bantam chicken named Peppa, trees, and grasses surrounded by wire so the chickens can't obliterate them completely and instead enjoy the blades as they grow from between the mesh. She has a garden of raised beds filled with kale, peppers, tomatoes, oregano, flowers, and lemon verbena. Her chickens are so loved, so intelligent, and so understood that Isabelle taught one to pick the queen of spades from a pack of cards. "It wasn't even that hard," she confides when I ask.

Chickens normally get a bad rap when it comes to intelligence. They're presumed to have small brains and incredibly limited capacities for emotion and cognition, but this isn't the case. As it turns out, the term *birdbrained* might not be such a seething indictment. A study published in the *National Review of Neuroscience* about the evolution of avian brains and mammalian brains showed that avian forebrains are derived from similar substances as mammalian forebrains—the part of our brain that helps us problem-solve and deals with complex cognitive abilities. This suggests that birds and mammals might have more in common, neurologically speaking, than has been previously assumed. Other research, published in 2017 in *Animal Cognition*, has shown that chickens possess the capacity to understand object permanence, employ deductive reasoning in social situations pertaining to pecking order, and resist the instant gratification of receiving a small amount of food sooner if they can choose to receive a larger amount of food later. A 2018 article in *National Geographic* referenced a study that showed chickens prefer beautiful humans and seem to value the same markers of physical attractiveness, like facial symmetry, that humans do. There's admittedly more research to be done to untangle the qualities of domestic chickens' intelligence, but it isn't surprising that Isabelle was able to teach her hen how to pick out the queen of spades.

The tour group gathers in a layered horseshoe of chairs in the shade at a far corner of the yard. Jan, a sensible-looking woman with tanned forearms and long fingers, and Isabelle are at the open end of the horseshoe. Jan works with a volunteer organization that rescues and rehabilitates chickens and holds a white leghorn named Sugar for a chicken health demonstration. About a year ago, Sugar was rescued from an egg farm; if she hadn't been, she would have been gassed during the normal "depopulation" process.

Laying hens are routinely replaced, and this isn't just according to Isabelle.

When the chickens are approximately eighteen months old and their productive abilities start to decrease, workers at egg farms will euthanize them with CO_2. Sometimes the depopulation process happens a little later, but the eighteen-month mark is a reasonable approximation; similar timelines have been reported on frequently by a variety of organizations. The US Department of Agriculture's *Poultry Industry Manual* cites that depending on manipulation of lighting cycles and limiting food to induce molting and boost egg production, the "productive life of laying hens" ends between 78 and 150 weeks. While it might seem like it would make sense to keep a commercial chicken for as long as she will lay eggs, this is not the case. Although they might continue to lay for almost three years, chickens are slaughtered as their productive abilities decrease, not when their productive abilities are void. According to the Food and Agriculture Organization of the United Nations, at 65 to 70 weeks, the percentage of birds laying will have dropped from a peak of 94 percent to 70 percent. In a flock of only 100 birds, this could mean a difference of approximately a hundred eggs per week. Manipulating environmental conditions will only make so much of a difference: older birds simply do not lay as reliably as younger hens. A new batch of laying hens replaces the old one to keep egg production high.

According to Isabelle, organizations are given permission to rescue a couple thousand chickens before the process begins, as long as the rescuers don't disclose the name of the farm to the public. The farmers like the agreement because it's cheaper; they save on the costs of labor and gas. And the volunteers like it because they get to save lives. Although it seems strange to have organizations devoted to rescuing hens from farming facilities, this is a widespread phenomenon. There are sanctuaries for farm animals in approximately twenty-four states and in other countries.

During the health demonstration, Jan tells us that a few drops of Frontline can clear up external parasites; ibuprofen can be crushed, added to water, and given to a chicken via oral syringe for pain; and sometimes you have to trim your chicken's toenails with dog nail clippers if they grow too long. She extends Sugar's wing to expose her flight feathers, clips them with a pair of scissors, and encourages us to feel comfortable doing these basic procedures ourselves.

Jan tells us that we can tell a lot about a chicken by its activity level, the color of its comb, and if it has any kind of discharge coming from its nose or eyes. Sugar goes *burrup, burrup* in her arms, and Jan feeds her cut grapes. "Know your chicken. Know who she is and what's normal."

Sugar gets passed from lap to lap, and it appears she's enjoying the attention. People massage her, pet her wings, and scratch under her chin. They marvel at how docile she is. Sugar croons and squints. When it is my turn to hold her, she feels like a warm loaf of bread in my arms: just as accepting, just as passive, just as soft. There's something about holding a chicken that's somehow uncanny. Chickens are odd creatures, somehow both prehistoric and cute, with their scaly legs, skinny heads, and small eyes. They run like I imagine bipedal dinosaurs designed for hunting might have. It seems like they wouldn't enjoy being held, these small miraculously domesticated dinosaurs, but they often do. I feel lucky when I get to hold a chicken, regardless of how often it happens, the same way I feel lucky when I get to hold a cat.

Toward the end of the discussion, Jan takes questions. Someone in the crowd asks her to clarify a comment she made about egg-laying hens being killed at eighteen months. It's clear that most people in the group have never heard of this practice.

"Get them in your backyard," Isabelle encourages. Isabelle suggests that people who are new to chickens get a coop that's larger than what they need, regardless of how many chickens they plan on adopting. "You will fall in love with them," she assures us.

CHAPTER 2

"Chicken math" is a relatively common term in the backyard chicken community. It's a little like eating Oreos—you *might* plan on only two or three, but soon you'll find yourself housing five, then seven, then fifteen. It just happens when you're not looking.

Isabelle knows this well: she's experienced a version of it. When she and her husband, Peter, bought their house nine years ago, they were vegetarian, wanted humanely sourced eggs, and had the space for a chicken coop. They decided that raising chickens in their backyard would be the best way to ensure their eggs came from hens who were treated well and cared for properly. They started with three pullets, young hens less than a year old.

About the time Isabelle began wanting more chickens, one of her friends became interested in getting a dog. Her friend went to a breeder on the East Coast, and Isabelle, who volunteered at a local animal shelter and was familiar with dogs who needed homes, was understandably upset. Her initial disappointment led to a realization: by buying chickens from a hatchery, she was doing the same thing as her friend who bought a dog from a breeder. The epiphany changed Isabelle's life.

Isabelle adopted two red sex-link hens from Animal Place, a sanctuary that, among other things, rescues chickens from egg factories, rehabilitates them, and puts them up for adoption. "The way they acted, they were so happy, so full of life, like they were enjoying their every moment to its fullest—that was amazing," she says. Isabelle's new chickens were surprisingly relaxed, took naps under the trees, and explored the garden. They got to act like chickens for the first time after a life of industrial agriculture. They ran after bugs but weren't sure what to do with them. When it rained, the rescued hens didn't care: they stayed outside the coop to explore. Because they'd been denied the ability to dust bathe for most of their lives, they spent hours doing it with gusto. Dust bathing is a process by which chickens dig and scratch into the dirt, creating a divot that they can get into. Shuffling from side to side, a chicken will use her wings to throw dust into her feathers, shaking them so the dirt penetrates to her skin. The process looks like a chicken's version of a pig wallowing or a sparrow in tray of water. Isabelle had to change the irrigation system around her trees in the chicken run to save it from their persistent claws.

Isabelle became inspired to volunteer with Animal Place. She learned about chicken health by providing care to the thousands of chickens brought home from each rescue: checking their vents, beaks, feet, and skin. She flagged chickens who returned from farms with broken wings and prolapsed uteruses and gave them delousing medication.

"They have tons of lice," she says. "It does not impact the egg production so farmers don't care." Isabelle learned about animal welfare and factory farms. After going on a rescue to an egg farm, Isabelle decided she wasn't going to eat the eggs from her backyard. "Even though I was not eating those eggs anymore, just seeing the way the hens in those conditions were treated just for eggs, it's like, it's not worth it, so I stopped."

Today most of Isabelle's chickens, besides Peppa and some grandfathered-in hens, were rescued from battery cages and taken from the farms shortly before they were scheduled to die. You wouldn't know it from the way they allow themselves to be held, scratch in the dirt, and socialize amicably with one another and the Tour de Coopers, but their malformed beaks and missing eyes ominously suggest a more traumatic past.

Before she went on a rescue, Isabelle wasn't sure she wanted to. She'd heard horror stories of finding body parts in cages. "I hesitated a long time before going on that rescue because it was like, do I really want to see? You know? But it's like, I have to see. I have to witness what is going on."

Isabelle and the other rescuers ventured to the farm in the darkness of a new moon. With a night so dense, the chickens were less likely to be active as they were removed from their cages, the only world they'd ever known. For them, the universe could be summarized by a few feet of wire mesh, a feeder, and a watering nipple from which they could drink. Isabelle smelled the barn before she saw it: the concentrated feces of the henhouse released ammonia into the California night air and produced an overwhelming stench that announced its presence like the olfactory equivalent of lightning before thunder.

Although this was no break-in—the company knew the rescuers were coming and both groups had clear knowledge of and consented to the events before they occurred—not all chicken rescues performed in the United States are coordinated. Not all are legal, or perhaps maybe one would argue that they are, but a strict sense of the word cannot be applied.

In 2018 a group called Direct Action Everywhere, or DxE, went into Weber Family Farms, an egg farm in Petaluma, California. DxE is an animal rights organization committed to rescuing animals, raising awareness about their conditions, and using nonviolent practices to expose the violence systemic to factory farms. Their goal was to remove sick or injured chickens from the premises and expose some of the realities of the farm. The five hundred or so demonstrators were not permitted to be on the farm and not given permission to enter the property, and Weber Family Farms had no prior knowledge about the DxE protest. After going into the facility and removing approximately ten chickens, the activists returned to the property despite a growing police presence. Forty activists were arrested. Later in 2018, DxE staged another break-in in Petaluma, this time at McCoy's Poultry Farm. Fifty-eight activists were arrested, one of whom was charged with seven felony counts, including felony conspiracy and felony burglary. Organizations like Animal Place, however, do not engage in illegal direct action and instead work with farmers.

When Isabelle arrived at the farm for the rescue, she saw that the barn was sided with sheets of chicken wire. The open-air design allowed the smell—maybe the most notable aspect of the operation—to escape, but it also allowed the weather to pummel the chickens. Most modern farms are enclosed, and for good reason. The bowels of these sprawling metal behemoths are unsightly. However, this one was older and wasn't enclosed. The industrial bulbs illuminated a long line of cages hanging from the roof, suspending chickens above the ground as if they were floating.

Contrasted against the darkness beyond the henhouse, the white hens caught the light. "The first thing I saw was those chickens hanging in the air," says Isabelle. "All those birds, kind of like little ghosts, hanging there. That was a really weird sight." The rescuers wore gloves and boots. Some wore masks but others didn't because of the heat.

In the barn ten thousand chickens were organized in three long double rows of cages. Each chicken, Isabelle tells me, was given approximately an iPad's worth of space. Their beaks had been trimmed, a process known as "debeaking," for safety. Debeaking is the common practice of removing a third to a half of a chicken's beak. It's accomplished several ways, but the most popular method in the United States is to use a hot (around 650 to 750 degrees Celsius) guillotine-type blade that simultaneously cuts and cauterizes. This is done without anesthesia on chicks who are usually five to ten days old. It's sometimes repeated later if the beak grows back.

Debeaking is a painful process, but it's instituted to avoid a much grimmer alternative. Chickens exhibit a number of behaviors like scratching in the dirt, dust bathing, and foraging. These activities are as central to the "chickenness" of a chicken as socialization is to the humanness of a human. When chickens can't be chickens, it stresses them out. Some chickens peck feathers, which results in painful wounds, but others start to cannibalize their cellmates. According to a 2009 report published in the Humane Society Institute for Science and Policy's *Impacts on Farm Animals*, poultry cannibalism has been reported in all types of housing systems and is a learned behavior. Chickens share this knowledge among the flock, and outbreaks are difficult to control. If debeaking were eradicated, chickens would use those sensitive, vital organs full of nerve endings to maim and kill each other in response to their unnatural and limiting housing conditions. G. John Benson wrote in *The Well-Being of Farm Animals* that "chopping off parts of young animals in order to prevent future welfare problems is a very crude solution," although it appears to be the one that the industry has chosen.

Some studies have shown that the drop in cannibalization is tied to guarding behaviors. Not only does debeaking reduce aggression (or at least, one of the aggressive outlets that birds with intact beaks would otherwise turn to) but it also changes the ways chickens interact with the environment. They peck, shake their heads, and wipe their beaks less. Although some people misguidedly think that debeaking is akin to a human clipping their fingernails (ignoring that a chicken beak is a sensory organ containing nerve endings), this instead seems to confirm that debeaking, as Michael Gentle and his fellow researchers at the Institute of Animal Physiology and

Genetics Research in Edinburgh, Scotland, described, "increases dozing and general inactivity, behaviors associated with long-term chronic pain and depression." In fact, the neuromas (areas of nerve regrowth and enlargement near the ends of damaged nerves) that can develop on the beak-stump post-trimming can have discharge patterns similar to those experienced by human amputees with phantom limb pain.

"They have nothing to do," Isabelle says of egg-layers in confined conditions. "They have no perch, no dust bath. It's like you're on an elevator with so many people for eighteen months. You get water, you get food, but that's basically it. They get super frustrated, they get crazy, they peck at each other." Instead of giving them more space, the industry standard is to debeak the chickens. Because debeaking only treats a symptom and not the cause, caged egg-laying chickens can resort to things like rubbing their breasts against the wire until they bleed. Mostly, it's out of boredom.

The chickens on the farm Isabelle visited, all things considered, were relatively lucky, if you can stretch the word that far. "The shed we went in had only one layer of cages; some had multiple tiers." Sometimes the cages are stacked up to five levels high. Below the cages was the source of the stench. The feces created a writhing sea, which was inhabited by maggots and rats. When Isabelle and the other volunteers waded through the muck to pull chickens from their cages, it sloshed into their boots.

Isabelle carried clippers. The chickens' nails grow long and sometimes bind them to the bars of their cage. "There are some of them that just die because they are stuck there and can't even reach the food." The chickens can become entombed by their long nails. "You have to cut them free to liberate them." Fortunately, Isabelle found very few dead bodies.

CHAPTER 2

The rescuers worked in teams of three. One person took the chicken from the cage, another opened the door of a crate to put the chicken in, and the third made sure the chickens already in the crate didn't escape. When one crate was full, someone took it outside and retrieved another. The process continued until there were no more available crates. The rescue was only meant to save fifteen hundred chickens, the maximum the sanctuary could accommodate and rehabilitate. "I remember the moment where I brought a crate outside, went to the pile where the empty creates were, and it was like, 'Where are they?'" Isabelle didn't realize they were running out of space until it happened. "I made the big mistake of turning back to the shed and looking at all the other chickens that were remaining. I knew they were going to be gassed the next day." The remaining chickens hung there like ghosts, seemingly lost. "You see that you can't save them all, that you're not going to save them all, and that's the lowest point of the rescue."

For Isabelle, the highest point came after transporting the chickens back to the rescue sanctuary. It's the moment the crates are opened and the birds freed, when they get to stretch their wings and touch dirt for the first time in their lives. Once in the open, the chickens immediately dust bathe, which has the effect of smothering their lice infestations. The rest of the rehabilitation process following their rescue takes about a month—volunteers clean the barn, feed the chickens, and make sure they don't have any health problems. The first few weeks are the most intense.

From the first night, volunteers are needed in the barn for a process known as *declumping*. Since the chickens have never been given space, they're afraid of it. Chickens can be naturally stressed by the darkness. They are prey animals, and to even my untraumatized chickens at home in Louisiana, dusk signals that it is time to find a protected place to spend the night. To the rescued chickens the barn is massive and terrifying in its sheer spaciousness, so they self-soothe the only way they know how. They ignore the roosting bars, foreign structures they'll only learn to use later. Since there's comfort in consistency and the hens are in an unfamiliar space, they try to recreate familiar situations. The chickens try to get as close as possible and form piles, masses that can reach ten hens high. These deadly stacks will crush the chickens on the bottom.

It can take two weeks for the chickens to understand that they can use the space, which means fourteen consecutive days of volunteers staying up at night, walking through the henhouses, ensuring the saved chickens don't accidentally kill one another.

I stop at a table with pamphlets and buy two bags of omega-3-rich chicken forage blend, which I plan to plant in my chicken run at home. I say goodbye to Isabelle, and one of the volunteers helping her gives me a free sample of vegan chocolate.

On the hour-and-a-half-long drive back to my parents' house, my mom's bike resting on the folded-down backseat, with front tire removed to make it fit, I think about the range of attitudes about chicken ownership that I witnessed on the tour. Trendy coops, $200 chickens, organic feed. All of these things beg several questions: are environmental movements just vapid expressions of wealth? Do people who care about the environment care more about their image as environmentalists? If people don't clean their own coops, are their chickens actually responsibilities or are they just cute lawn décor that someone else maintains? Is it easier to resist corporate agribusiness if all you have to do is write a check? And finally, most

CHAPTER 2

importantly, if people couldn't just write a check, if having chickens wasn't something that would grant status, if it necessarily involved real labor, contact with feces, and sweat, and if those things couldn't be outsourced, would people still decide that it matters? Would they feel so passionately about healthier homegrown eggs, about the atrocities of modern agriculture, or about connecting with the land?

Yes. No. Maybe.

These are all fair responses; there isn't one type of chicken owner. It's inevitable that some people would never consider owning chickens if they weren't cute or if the coop wasn't aesthetically pleasing. It's also inevitable that some people would always choose cleaning out a coop over a fancy dinner, every time, hands down. The coop tours represented the spectrum. While some coop owners sipped white wine in expensive-looking linen rompers and straw hats, others were in jeans, covered in straw, shoveling dirt and compost into the garden, clearly implying what everyone who actually engages with these types of behaviors knows so deeply: the work never ends, the chickens always need something, and there is always more to do.

3

Urban Agriculture

WHEN CHICKENS COME TO TOWN

THE CHICKENS WATCH me while I squat before the garden: they know what the position alongside this blue bucket means. I pick bugs off of the plants and toss them in, occasionally adding earthworms, although the garden could use their help. When I'm finished, I pick several cucumbers, their vines twining around a strange amalgamation of wire, stakes, and fragments of an old coatrack. These vines are vicious and constantly threaten to overwhelm the tomato plants—given only a few days of unsupervised growth I'm sure they would seize the opportunity. The garden is small, a miniature rectangle dug into the corner of the lawn, but it's productive enough to yield a half year's worth of pickles, a few pounds of fist-sized tomatoes, and a handful of peppers. My vining sweet potatoes, wild-growth cucumber bushes, leggy tomato plants, and backyard hens are all examples of urban agriculture, although sometimes they seem a little wilder than domestic. I walk over to the chicken run and tip the contents of the blue bucket onto the ground. It's only a few palmfuls of bug protein, but as they descend upon the writhing bugs, the chickens are fierce. Grendel grabs the largest worm, thwacks it against the ground, and tears it in half. Amelia snags a piece and runs away. I've heard about people collecting snails, purging them, and frying them with butter for dinner, but watching the chickens attack the invertebrates doesn't make me salivate. Once they're done with the slaughter, Amelia wipes her beak on the grass by dragging first one side and then the other. She shakes her head and goes to the other end of the yard to scratch at something in the dirt.

Urban agriculture is the process of using city spaces—whether they be backyards, abandoned lots, or rooftops—to grow food, rear animals, and sometimes support communities through land engagement and access to fresh produce. The United Nations Development Programme estimates that approximately 100 million people worldwide earn a portion of their income from urban agriculture. Research published in *Agronomy for Sustainable Development* has suggested that as the world becomes increasingly urbanized on a global scale, the amount of urban and peripherally urban spaces devoted to growing food and raising animals has also increased. This could be seen as a desire to touch nature, either in spite of or due to the increasing conglomerate of glass, concrete, and steel that humans encounter on a daily basis.

One of the theories behind the rising fascination with urban agriculture is that urbanization and capitalism alienate us from the natural world. Urban farms, while existing within these systems, salve this increasingly chronic, and arguably infected, wound. Nathan McClintock, a geographer and professor of urban studies and planning at Portland State University, said that urban agriculture is becoming more popular because it combats some of the shortcomings of our consumer society. We consume food that is commoditized and live in places that have lost their natural biospheres. "By rescaling production, reclaiming vacant land and 'de-alienating' urban dwellers from their food," writes McClintock, urban agriculture "attempts to overcome these forms of rift." He sees urban agriculture as something that can help people find belonging in a way previously denied by capitalism.

Like the backyard chicken movement's reclamation of what's appropriate for modern city spaces, urban farming challenges the notion that food production should take place elsewhere. It disrupts one of capitalism's grand narratives: that there is a massive rift between production and consumption, and what you produce is meant to be sold to someone else and that it's not *really* yours.

Under a capitalist system, it's not very common to consume the things we produce or to produce the things we consume. On a very basic level, capitalism works through the private ownership of capital goods, or goods that are used for the production of things that consumers will use, known as consumer goods. It's a profit-driven system in which those who possess ownership of the buildings, machines, tools, raw materials, and everything else that goes into the production of a consumer good have an incredible degree of control. Workers sell their labor for wages: they "clock in," so to speak, and use things that are not their own, like pens and rubber and iron and oil, in combination with whatever skill or time their job requires, to produce something for their employers. Although workers' wages represent a trade of sorts, it's not exactly fair, because to continue employment a worker must always earn more money for the company than what they are paid in wages. Why else pay them? If a shoe company spent more on rubber than they could sell shoes for, they wouldn't buy that rubber anymore. In the same way, if a worker's wage is more than what they make for the company in the time they are paid to be there, the company wouldn't employ them anymore. Businesses aren't employing people out of the goodness of their hearts; they're doing so to make money, and if a worker cannot ultimately earn more for the company than they require the company to pay them, then there's no reason for the worker to be there.

Working for someone else has a variety of effects. If I'm a poultry worker on a processing line making just one or two select, repetitive cuts to a pre-cleaned carcass every day, I probably don't have a strong sense of ownership over the final product that ends up in a grocery store. I wouldn't be proud of the chicken because I didn't produce it; I just made a couple of cuts. Imagine, instead, that I'm an independent farmer and have raised a chicken, slaughtered it, and am then viewing the cleaned product ready to go to a consumer. Not only did I have a lot more control over the process but I would probably feel a sense of earned pride. When people can't apply individual creativity or self-direction to their work, they can experience a sense of alienation from the final product.

People who do the work of production—the employees who build things and make food, even those who provide abstract services like data analysis and computer technology—don't generally decide what they'll produce. Sure, people can choose to apply for one job over another, but once they're hired their capacity to direct their labor and choose what they want to do on a day-to-day basis is limited.

The ability to directly consume what you produce and control the conditions of your labor—when it takes place, how it takes place, for how long it occurs—fundamentally shifts the relationship people have to their work. There's the possibility that, through certain forms of urban agriculture or self-directed work, people can reconnect to their labor rather than working for the sake of accumulating wealth.

Our roles as consumers and producers in a capitalist society are slippery, but the ability to either own what we produce or to freely direct our labor according to our own wishes articulates a different kind of relationship than that exemplified by institutional production. This isn't to say that there aren't benefits to institutional production. I have no idea how to fix a broken transmission or a leaky toilet. I don't have the desire, let alone the time, skill, land, or tools, to mill my own flour or grow my own oats. I don't want to spend time making all of my own clothes, blankets, hair ties, dishes, pots, and pans, nor would I be able to. If I were to only consume what I myself produce with my two hands, my world would become much smaller and impossible to manage. I wouldn't be able to read books or own a cell phone. It's impossible to imagine an existence in which I only relied on my own resources and abilities, but I can imagine an existence in which people have more control over certain aspects of their lives and a closer understanding of what kinds of goods, practices, and labor their existence requires. Closing the gap between what we produce and consume is impossible, but at the very least an appreciation of what production requires gives people some of the tools they need to be responsible consumers. It helps create an understanding of how we want to engage with and occupy these roles.

Although it would be possible for me to buy eggs from a store, I'd be trading money, an unintelligible product removed from my labor that produced it, for goods. When I raise chickens in my backyard, there's a direct relationship between my care for them, or my labor, and the products I get, like manure for the compost and eggs for breakfast. I don't have a sense of estrangement from what I consume because I had a direct, personal, and clear relationship in the conditions of its production.

CHAPTER 3

Backyard chickens are a counter-narrative to the idea that, in a capitalist society, individuals occupy separate spheres when they are acting as producers who earn money for others and when they act as consumers who trade money for goods. Backyard chickens reorganize these roles and modify them so people work for themselves in a new, small way. While the zeitgeist of the late nineteenth and early twentieth centuries sought metropolitan spaces and housing that were markedly different than production spaces, scrubbed clean of animals and what society thought they stood for, the United States now faces an increasingly hyper-technical, hands-off world where that separation between production and consumption often seems absolute. It is quite possible to eat food without ever growing it and to spend a lifetime consuming exclusively factory-farmed eggs, meat, and dairy. In fact, most of us do. Backyard chickens and other forms of urban agriculture purposefully reintegrate ownership of and contact with food production into people's lives.

Nicole Graham and Chicken Mike are the married masterminds behind The Garden Hen, a Houston-based urban chicken company that supports people interested in starting or maintaining their own backyard flocks. The Garden Hen builds coops, breeds chickens, provides all types of poultry care, fights for inclusive backyard chicken laws, and works to educate people on how and why to raise chickens. "We're a one-stop chicken shop," Nicole told me one afternoon when I visited her and Chicken Mike in Houston.

Nicole is a self-described "toes in the dirt kind of girl" who climbed trees as a kid, is a tomboy a heart, and was slaughtering and cleaning quail—a previous business venture—when she was four months pregnant. She's in her midthirties and has sleek dark hair, a lean build, and an electric smile. Chicken Mike looks like a salt-of-the-earth man with a charismatic flair that makes him suited for the public-facing nature of their business. He unironically wears a cowboy hat, has short facial hair, and possesses an assortment of tattoos, including an outline of Texas, a rooster, and Texas bluebells.

Chicken Mike and Nicole didn't always have chickens, and when Nicole initially proposed that they start their own backyard brood, Mike (it was just "Mike" back then) was scared of birds. It was a childhood fear; when Mike went hunting as a kid, he refused to put dead birds in the quail pouch on his vest. He was too scared they'd come back to life. I imagine hearing his wife wanted to bring chickens home was somewhat nightmarish. Mike had one stipulation: "It's all you," he told her and made a hard pass on even touching the chickens.

Nicole built her own coop and fortified a dog run. She fed the chickens, watered them, collected eggs, and repainted the coop every year, without Mike's help. "You have to say, 'Okay, I'm committing,' and fully throw yourself into it," Nicole said. "That's exactly what I did. I said, 'Okay, this is where I'm going with this.'"

At the time, Nicole was an assistant teacher at a private school with a hatching program. "I started volunteering to get the eggs, and then I moved from volunteering to get the eggs to setting all the incubators, then from that I basically just took it all upon myself." Eventually Nicole told herself, "Well, I'm just going to take all the babies home." After all, they already had the coop.

It wasn't until about three hatching programs in that Mike caught on. Remember chicken math? It's the common phenomenon of getting just two or three chickens, then waking up one day and realizing that your yard is full of hens, you need a bucket rather than a bowl to carry feed, and the local feed store's springtime chick stock will still make you breathless.

The question hit Mike all at once: "Why do we have fifteen chickens running around?"

CHAPTER 3

Even as their yard reached its chicken carrying capacity, Mike still wasn't enraptured with them. Anytime a chicken needed to be caught, transported, or examined, Mike would feign helping. He'd act like he'd try to catch it while ensuring he never came close to touching one. He'd imitate a grab, Nicole would snag the bird, and Mike would save face by saying something like, "Oh darn, I missed it," or "I shushed it over to you, why didn't you grab it?"

It was almost surreal hearing the Grahams talk about their early days of chicken-rearing. Fast-forward to the present day and Mike is called Chicken Mike. He handles chickens every day. As he hauls pine shavings for a client's coop from his truck, I watch the rooster tattooed on his arm move with the muscle.

Pinecone, their self-proclaimed therapy chicken, changed everything for Chicken Mike. Pinecone is a six-year-old frizzle, a type of chicken whose most remarkable feature might be their feathers. Instead of folding inward and back in a sleek mat, the feathers of frizzles fly outward and forward: they look like they've been brushed the wrong way. With her brown coloration, Pinecone actually looks like one. She helped Chicken Mike get over his bird-related fears.

It's not uncommon for someone to walk up to Nicole and Chicken Mike—who bring Pinecone everywhere, from bars to schools to the state capitol during an effort to lobby for increased allowances for backyard chickens—and say that they would never touch a chicken. But Pinecone is too charming. "Before you know it," Mike says, "the chicken is on their shoulder."

Pinecone often participates in their outreach and educational programs, which are a vital part of The Garden Hen's existence. When I ask Nicole why this aspect is so central to their business and why spreading awareness about this type of urban agriculture is so important, she tells me, "The reality is deeper than just a chicken in your backyard. It's connecting with nature. It's passion. It's love. It's an animal that gives back." Urban agriculture can fight policy failures like lack of infrastructure and even institutionalized racism and classism.

When I asked about why they felt so strongly about promoting backyard chickens through their business platform, Chicken Mike told me, "We have the third largest food desert population here in the United States. Houston does." A food desert is an area with limited access to affordable and healthy food, like fresh vegetables and fruits. Supermarkets are few and far between in food deserts; instead, these spaces are often populated with mini-marts and convenience stores that stock their shelves with highly processed, sugar-rich, and fat-laden foods.

The term *food desert* would have been impossible in a bygone era when animal agriculture was a fundamental aspect of city infrastructure: hogs consumed household waste, chickens pecked in yards, and goats grazed. During the early twentieth century, when zoning ordinances banned most livestock, productive animals were increasingly removed from urban spaces. Animals' existence in cities was once central to food security, and their elimination was often targeted at low-income and immigrant neighborhoods and families. Even now, because of lot lines and wealth, the capacity to raise animals like chickens does not apply universally to all residents regardless of class.

Food deserts mainly exist in communities of color and low-income areas. Studies have shown that predominantly white neighborhoods have, on average, four times more grocery stores than predominantly Black neighborhoods and that the grocery stores in white neighborhoods are larger and have more healthy options. The relationship between food availability and racial and socioeconomic segregation of neighborhoods is staggering.

CHAPTER 3

Proponents of urban agriculture, like Chicken Mike and Nicole, are in a pivotal position to work against oppressive food structures. While the Grahams started a business with various educational, social, and activist branches, others, like Kathleen Blakistone and Richard Draut, wanted to *simply* start their own urban farm.

Moonwater Farm, a micro-farm in Compton, California, has been actively working to undo some of the power structures that contribute to food deserts and disenfranchisement for years. The farm hosts workshops, classes, and custom-designed field trips in organic farming, livestock raising, healthy food preparation, permaculture, woodworking, and soil regeneration, among other topics. In Compton the murder rate is four times the national average. Gang violence and drug abuse are historically pervasive problems. Compton has high crime rates, high poverty rates, and some of the highest diet-related disease rates in all of Los Angeles County.

The micro-farm was originally meant to be an aquaponic operation, a system that combines raising fish and then producing plants in nutrient-rich water. When Kathleen and Richard bought their suburban lot, they envisioned restoring the worn house at the front of the property and using aquaponics to grow lettuce for high-end restaurants. On the first night in their finished house, a cowboy inexplicably knocked on the front door. "He asked us if he could lease our backyard for horses," Kathleen told me.

Cowboys, let alone horses, were not part of the original plan. Kathleen had envisioned fish and leafy greens, not livestock. "We said, 'Well, we're going to have this greenhouse, so probably not, but we'll call you back.'" Richard and Kathleen thought about it, changed their minds, and spoke with him the next day. "Listen," Kathleen said, "if you want, you can come for six months, but we're going to build this."

Kathleen and the cowboy connected almost instantly. They talked about urban farming, agriculture, and what it would mean for the community if children had access to healthy food and people had access to land. He brought Boy Scout troops to the backyard lot to see the horses, and Kathleen made healthy, plant-based snacks for them. They taught kids about agriculture and the importance of food. At some point it became clear that the backyard was not going to, and should not, exclusively house leafy greens. The idea of an aquaponic farm faded as something more dynamic emerged. "We just realized it was going to be a lot more interesting growing people than lettuce, so we took a left turn," Kathleen said.

"We are in Compton, which is a historically marginalized community for a whole lot of reasons," Kathleen told me. The backyard chicken movement is often predicated on the idea that people have access to land, whether that space is a cramped lawn between lots or something expansive in the country, but systemic racial inequality is a reality of America's land distribution. From the 1930s to 1960s, the Federal Housing Administration institutionalized racism by granting mortgages to white home buyers, making it possible for many white families to own a home for the first time, while excluding non-white families from those opportunities. The FHA deemed non-white families credit risks, refused to lend them money, and enacted a process of "redlining" that all but denied basic financial services in predominantly non-white communities and encouraged segregated neighborhoods. Today, the effects are stark: white families have almost ten times the net worth of Black families, and the homeownership gap between Black and white households ranges from 14 to 50 percent, depending on location.

Homeownership, access to outdoor spaces, and the ability to connect with the land are enmeshed with one another. Moonwater Farm

CHAPTER 3

recognizes this. Accordingly, Kathleen says, "Our goal is to provide access to land and land-based activities for folks who traditionally don't have that. We've been here for several years now and done work in the community and work understanding white supremacy culture."

Access to land in the United States has always been a fraught topic. The entire homesteading enterprise in the early history of the United States was a project based in white supremacy and involved removing Indigenous people from tribal homelands, genocide, and enslavement. While the food movement can help negate white supremacy and racial capitalism in certain ways, it must simultaneously recognize how it traditionally has been bound to those systems. There can be a tendency to create an idyllic picture of a bygone era when everyone ate the food they grew, but the realities of land-building and personal agriculture were not so, and the capacity to rewrite history comes from a place of privilege. In the wake of all of this, the work individuals are doing to counter the ties white supremacy and land access have is even more important.

"We are deeply committed to the idea that we need to regenerate our soils and our communities in order to become more resilient and survive the challenges that are facing us," Kathleen told me.

Like parts of Houston, Compton can be called a food desert, although some prefer a different term. "There's essentially a food apartheid that's evident in our urban spaces," Kathleen said. "That's by design because we redlined people of color not to live in certain places, and there's no grocery stores or fresh food or education systems serving those young people." Activists like Karen Washington have argued that *food apartheid*, rather than *food desert*, appropriately references the fact that social inequalities and injustices are linked to the systems that designate who does and doesn't have access to healthy produce, grocery stores, and land. In a 2018 interview with the *Guardian*, Washington said that the term *food apartheid* "looks at the whole food system, along with race, geography, faith, and economics. You say 'food apartheid' and you get to the root cause of some of the problems around the food system." In the article, Washington explained that the term *food apartheid* "brings us to the more important question: what are some of the social inequities that you see, and what are you doing to erase some of the injustices?"

The word *apartheid* means "apartness" in Afrikaans and denotes the system of legislative work that enforced segregationist policies against non-white South African citizens. One of the reasons the term *food apartheid* is perhaps more fitting than *food desert* is that it avoids the naturalistic language of the latter term. It's not a natural, land-based metaphor. There's nothing *natural* about a food desert; it is the result of systemic oppression and dynamic man-made forces.

Unfortunately, part of the problem isn't always access. In 2019 the *Quarterly Journal of Economics* published a study, "Food Deserts and the Causes of Nutritional Inequality," that revealed grocery stores won't necessarily fix the problem posed by food apartheid. At the heart of the issue is the finding that individuals in low-income households consume diets that are less nutritious than individuals in higher-income households, a trend that results in low-income communities experiencing higher instances of nutritional, health, and dietary issues, which can have lasting effects and result in high health-care-related costs and a deteriorated quality of life. The study found that offering low-income households access to grocery stores with healthy food options only reduced the trend of nutritional inequality by 9 percent. To put it another way: more access to grocery stores isn't a cure-all to areas affected by food apartheid. New supermarkets in low-income communities had a small effect on consumer purchasing habits.

CHAPTER 3

If you are someone who is working multiple jobs and cannot afford high food bills, it doesn't matter if there is a grocery store down the street. Under a strict food budget, what becomes important is getting the most calorie-dense foods for your dollar. Even if a grocery store with healthy options shows up, if food scarcity is an issue, those options can't change consumer habits if you can't afford them. The differences in organic versus non-organic produce or pasture-raised versus caged eggs become moot; what matters is being able to put food on the table, and access to grocery stores doesn't change the realities of low-income households. The main benefit, the study outlined, of putting grocery stores in areas with limited access is that it reduces travel time and transportation costs to go shopping, but 91 percent of individual households didn't experience a shift in their consumer habits.

That's one of the reasons why, as Nicole said, "it is deeper than a chicken in your backyard." A chicken can be a form of activism, one that Chicken Mike and Nicole know well. A lot of the children they teach live in areas suffering from food apartheid. "We teach them this skill of cultivating an egg in their backyard. We teach them something they can do on their own." The Garden Hen goes to different schools with their specific education program that raises awareness about backyard chickens and teaches children to care for them. According to Chicken Mike, their program combines "urban farming, backyard chicken farming, and adaptation skills for chickens." Nicole and Chicken Mike estimate that they have reached over sixteen thousand students in the Houston area.

In addition to education, they work from an activism angle to fight restrictive municipality ordinances and policies that keep chickens out of urban yards. "The City of Houston has a very inept hundred-foot rule," Chicken Mike told me. This rule states chickens must be kept one hundred feet, in any direction, from any neighboring properties. "So there are kids that live in the food desert, and they have to be a hundred feet from their neighbor to have a chicken in their backyard." Chicken Mike told me that areas where poverty is more rampant tend to have smaller lot lines, which limits individuals' capacity to raise chickens if they wish. People with larger lots in more affluent areas don't face the same restrictions. "It's just not a fair scenario," he said.

Although chickens are not for everyone and require a degree of disposable income (to afford their shelter and continued food requirements), the limits on chickens and systems that prevent individuals from owning them should not be compounded by regulations that target specific areas based on income and lot lines. Rather than picketing to enact change, Chicken Mike and Nicole petitioned the city through formal channels. They spoke at the Texas Senate on behalf of reevaluating the laws governing backyard chickens. By decreasing the availability of urban farming and urban agriculture, restrictive mandates disallow certain kinds of homegrown protein. "Our whole goal is to approach them through an education means and say, 'Hey, this is what we have been teaching kids and this is why it's smart to change this rule.'" While I was in Houston and visited urban coops, I noticed that this rule wasn't always honored, which reinforces the idea that chicken people are somehow illicit or counterculture, bending (and occasionally breaking) rules to keep hens near their homes.

In the past two decades, the validity of some of these mandates has been called into question. According to a survey conducted by the Trust for Public Land of America's hundred largest cities, between 2017 and 2018 the number of garden plots in city parks increased by nearly 22 percent to twenty-nine thousand nationwide. It's estimated that Americans spend $1 billion annually at farmers' markets. Some, like Kelly Coyne and Erik Knutzen, authors of how-to books on urban homesteading, believe that society will become more accepting of private property, city-dwelling urban livestock because the current industrial system is "untenable," and if people raise their own chickens, they'll know exactly where their food came from and how the animals were treated.

The idea of *relocalization*—of creating smaller food, energy, and production systems that people have more control over—is part of a global movement that fights the narrative that says some spaces should be urban while others industrial. This is a grand capitalist narrative and it separates the spaces where people can consume goods versus where the goods are produced. Urban agriculture challenges the notion that some people are only consumers while others are producers and recognizes that this myth denies a lot of communities access to fresh produce and healthy food. The places that are generally impacted the most, like Compton and areas of Houston, are the ones that have been historically disadvantaged by racism, redlining, and systemic oppression. It is unlikely that any system that denies choice on such a fundamental level as food doesn't have other critical ramifications for personal agency and individual opportunities.

Darren Chapman knows about the social ramifications of systemic issues. He is CEO and founder of Tiger Mountain Foundation, an organization that builds gardens in vacant lots to empower communities and combat the high recidivism rate in South Phoenix, Arizona. Tiger Mountain Foundation employs individuals who have been previously incarcerated and gives them the real-life and career skills necessary for rebuilding their lives and their communities. "You have a convenience store across the street from a convenience store across the street from a convenience store," Darren told me. "All of them have the best deals on cigarettes, the best deals on alcohol, the best deals on processed meat, the best deals on processed everything." There are no best deals on organic carrots at the local 7-Eleven, no clearance sales of wild-caught salmon.

This isn't just a South Phoenix problem: it's a national problem. A 2002 study published in the *American Journal of Preventive Medicine* examined the correlation between neighborhoods' socioeconomic levels, racial segregation, and access to food stores and service locations. There were four times as many supermarkets in predominantly white versus predominantly Black neighborhoods. The poorest neighborhoods had three times the number of places to consume alcohol as the wealthiest neighborhoods. Supermarkets in nonminority and wealthy communities not only stocked more food options than gas stations or convenience stores but often offered food at a variety of price levels and provided greater access to healthy food and food choices.

"You look at most of our corners and there's processed food and other different things that, we found out, will actually ail you," said Darren. "We have all the sugar, all the tobacco, all the alcohol. We are the number one consumers in the poorest community, so therefore our outlook is a bleak environmental outlook."

That bleak environmental outlook has tangible consequences, and Darren doesn't seem to see food access and social inequality as separate issues: "Here we are in the highest recidivism area code with all the gang trappings and fast food and other issues of disparity because it is disparity."

Urban farms can offer a way out. They generate jobs and income. They can teach people agricultural skills upon which to build a career. Tiger Mountain Foundation teaches people financial literacy, how to run a farmers' market booth, and how to grow produce and give back to the community in a tangible way. As a child, Darren's grandparents created a community garden, and Darren saw that growing fresh produce, sharing food with neighbors, and meeting people among the rows to swap skills created much-needed community conversations. The gang violence that occupied the streets was held at bay in the garden. His grandparents grew avocados, peaches, plums, greens, and carrots, drawing people from all over the community together. His grandfather taught himself Spanish so he could speak with and trade fruits and vegetables with his neighbors from South and Central America. The inclusivity exemplified by his grandparents and their garden was inspiring. "They could bring in thought processes that could integrate, initiate, and activate human beings into positive movement. That positive movement on an economic level, on an environmental level, on a built environmental level, actually becomes the initiative."

Now, decades later, Darren still understands the value of his grandparents' urban garden and recognizes the strength and empowerment that agriculture can bring into communities. By creating gardens on abandoned lots, Tiger Mountain Foundation rewrites the narrative of places disrupted by poverty. "One of the things that was established early on [at Tiger Mountain Foundation]," Darren told me, "was to change those spaces into something else. Like my grandparents' backyard—when I thought about it, it just made perfect sense that it was my salvation."

Tiger Mountain Foundation has always been driven by community. It's supported by neighborhood volunteers and gives ex-inmates the work experience, financial skills, and personal development to stay out of prison. The community support is threefold: people, places, prevention. Although the recidivism rate in South Phoenix is staggering, the rate of recidivism for those who go through Tiger Mountain Foundation's program is only 15 percent.

Darren described the first steps of transforming abandoned lots into community gardens to me. He starts by testing the soil to learn how to regenerate it. "I find that some pollutants are actually literally just falling from the sky, so now the soil culture is different after decades of this stuff falling on the ground. The soil has a certain type of content—that's from Big Brother, *thank you very much*—but as you dig through that soot and that layer of two, three, four inches, you start finding crack vials and 40-ounce bottles and other paraphernalia." Over time the soil is polluted by industrial influences, but underneath that layer of systemic waste there are products with individualized ramifications. "Welcome to your wasteland that we're calling the good ole United States of America," Darren added.

Darren seems to hold no illusions about corporations that maximize profits at the expense of individuals, and neither do Chicken Mike and Nicole. It seems that the more invested people are in small-scale agriculture, the more they're aware of some of the issues associated with the capitalist impulse to put profits above people.

"Big Chicken will be like Big Pharma," Chicken Mike told me. Big Chicken refers to the large industrial companies that profit from raising and selling chickens in CAFOs, or concentrated animal feeding operations, where broilers survive for seven weeks alongside twenty to thirty thousand of their kin. He says that the market won't shift until consumers demand it and compares this to the pharmaceutical industry: "Big Pharma doesn't want to make money selling cannabis as medicine. They already make money selling medicine as medicine." The systems that hold these agribusinesses together are already in place: housing sheds, processing facilities, chains of supply and demand for their meat, feathers, and even their soiled substrate (known as "poultry litter," this cheap by-product is often sold as a low-cost protein-rich feed supplement for cattle). Just like pharmaceutical companies are invested in the status quo because that's what's making them money, industrial agricultural operations that already make money selling conventional chicken aren't compelled to shift toward more local, sustainable alternatives until it positively affects their bottom line. "Big Chicken won't get behind backyard chickens and urban farming until—somehow—it can make them money."

The global agribusiness system is a staggering monument to capitalism's power and a by-product of predatory capitalist practices. Corporate agribusiness practices tend to devalue quality-based measurements in favor of quantity-based. Things like sustainability, worker happiness, environmental responsibility, healthfulness of food and systems aren't necessarily calculated as carefully, or deemed as valuable, as things like output, profit, and subsidies. It's hard to put worker happiness or environmental responsibility as a bottom line, but it's easy to put numbers concerning growth and sales there.

This isn't meant to say that our food system is a rational system; in fact, in some ways it's horribly irrational. For example, most countries don't refrigerate eggs, but American businesses do. When eggs are washed, a natural protective coating produced by the chicken, called the cuticle, is removed, and because eggs have pores, bacteria can be pushed through the shell. Because of this, washed eggs need to be refrigerated. In Europe it's illegal for companies to wash eggs; in America it's mandatory.

The rationality that backyard chicken keepers learn is an old form of knowledge previously denied to many. Through chickens, people can rediscover fundamental skills and a basic understanding of how the world works. Let me say this another way: our great-grandmothers would be just as baffled at the prospect of refrigerating eggs as some of us are by the idea of not refrigerating them. And no one would have needed to remind our great-grandfathers that eggs—like strawberries—are somewhat seasonal commodities. Without climate-controlled henhouses illuminated by industrial lighting, egg production would drop off in the darker, colder winter months.

The understanding that some things are seasonal is inherent to those who work with plants and animals. Shonda from Sbeck Traditions Backyard Homestead, located in Maryland, told me that in addition to eating fresh from the garden, her family "also preserved our harvests to feed ourselves after the growing seasons were done." Although her family wanted to buy a home with more land, they "kept getting outbid." Instead, they live in a private community in the woods, where some of their neighbors have chicken flocks. Shonda and her family are "doing the absolute most" with the space available to them. Their lot is only seventy-five thousand square feet, but they have ducks, quail, chickens, meat rabbits, and a garden. "Raising animals for meat seems like second nature to me," Shonda said. "I picked it up like it was something I've been doing all my life."

Urban agriculture provides a particular kind of intimacy in food production. "I wanted to be less dependent on the grocery store," Shonda told me. "I wanted our children to learn how to provide for themselves." Individual operations that allow people to feed their families and teach their children provide autonomy in a way that grocery stores can't.

Systems of agriculture that are small, personal, meaningful, and community-minded are exemplified by urban gardens that provide local food and land access. But they're also somewhat incompatible with capitalism because they almost always intend to serve communities first (and shareholders never) by holding people above profits.

At Moonwater Farm, Kathleen and her community "are exploring ways of potentially moving into more cooperative management and really trying to demonstrate alternative ways of exchange and conversation instead of pushing the consumption model." Rather than constantly buying and selling, Kathleen looks at barter and trade. Classes are offered on a sliding scale, furnishing the community with more access and the farm with fewer profits. "It's not a hugely profitable model, but that's not necessarily the point," she says. Finding what's important—what holds more human value than money and stands outside of a traditional large-scale capitalist system—has inherent value. Urban farms that do not seek to maximize profits over organic practices or community support and who want to, as Kathleen said, grow people instead of lettuce, choose communities over money. They undermine capitalism by illuminating other ways food can be produced and neighborhoods sustained. "We really look at how we can potentially work in smaller ways," Kathleen says. "Not larger ways, but smaller ways, which is often not a very capitalist approach." When operations and production systems are smaller, they can exist in the human gaps beyond capitalism and afford to value things that are qualitative, like community, equality, and empowerment—rather than quantitative, like profit margins, bottom lines, and overhead costs. Although Moonwater Farm and the type of agriculture it practices are quantitatively positive (studies published in *Science*, the peer-reviewed journal of the American Association for the Advancement of Science, and in *Sustainability*, an international scholarly peer-reviewed journal, have shown that organic farming improves soil fertility, has the capacity to sequester more carbon in soils than conventional agriculture, and contributes to diverse microbial activity that can stimulate plant growth), they still find space for qualitative values. Moving away from perpetual growth models means fostering what you do have and coveting your current resources rather than additional ones.

Buying into the relocalization movement implies moving away from large capitalist structures and investing more in your community. It means recognizing what is real given the parameters of environment, animal welfare, and community values. Eggs are seasonal and don't need refrigeration. Yolks are more orange than yellow. It's not rational to value money over people. Choosing qualitative values doesn't mean forgoing quantitative ones.

Sometimes urban agriculture is something smaller than an entire farm. It can just be a backyard. Alycia is a bubbly woman who laughs well and speaks with her hands. She lives with her husband, Kevin, in San Francisco, where they raise chickens for eggs, grow vegetables, and breed rabbits for meat. She has black, chin-length curly hair, which makes her seem practical, but there's also something undeniably accessible and playful about her. She jokes about the absurdity of her husband bringing home a giant rooster who needed rehoming because he hated women (when Alycia, a woman, works from home) and recognizes the particular and incredibly familiar temptation (at least, it's one I'm familiar with) to buy overpriced things that you don't need at Whole Foods ("Like an eighteen-dollar jar of peanut butter. It's like, 'NO! STOP,'" she says, eyes widening before she breaks into a laugh).

When Alycia met Kevin, he already had the house they live in together today. There were two or three chickens in the backyard.

CHAPTER 3

Kevin spent the previous year in graduate school at UCLA, and the roommate he kept in charge of the chickens wasn't particularly interested in his role as urban farmer. "They were in this old dog crate, swimming in mud because it was at the bottom of a hill, and he didn't have good drainage. Imagine a bachelor pad but with chickens." Alycia told me that everything was "in shambles," and that the first time she came to Kevin's house she remembers thinking, "I could totally fix this. I could do this way better."

Although Alycia grew up in Kentucky and knew a lot of people with chickens, at the time she'd been in San Francisco for about eleven years. Living in one of the city's apartment complexes, she didn't have access to outdoor space when she met Kevin. "I was craving gardening. My mom has always had a garden and I grew up with gardens, and Kevin had a greenhouse at the time and these chickens. It was another big selling point for going on another date." Things snowballed. "As we got closer and more serious and started making decisions together about what we were going to do on the farm and in the garden, we were just really dedicated to this lifestyle that we wanted to build around growing our own food."

Alycia favors fancy chickens: in her yard she keeps frizzles, a type of chicken with feathers that grow out and up; Ayam Cemanis who are entirely black, down to their meat, bones, and organs; silver-spangled Hamburgs with feather patterns not unlike the spotted markings of Dalmatians; and even naked necks, whose name implies exactly what it seems to: an entirely featherless neck.

Chicken breeds are somewhat like dog breeds in that both are the result of specific breeding and a lot of human selection for the sake of particular characteristics. A breed is an artificially selected group of animals: new breeds of chickens can be created over time, much like labradoodles were created. The term *species* defines animals of the same group who can successfully produce offspring who also breed. There are hundreds of chicken breeds out there, defined by things like their plumage, number of toes, combs, feather texture, skin color, and egg color. Alycia's flock represents a colorful mismatch of fantastical breeds with different characteristics and personalities.

Since many of the birds Alycia gets don't come sexed, a number of roosters are inevitable. As she tells me this, she holds a chicken named Yolandi, named after one of the lead singers of Die Antwoord. Yolandi, probably one of the few of her kind, is a two-toned cross between a frizzle and a blue laced red Wyandotte. These Wyandottes have orange and blue-gray feathers, making them look like walking homages to Van Gogh and his iconic use of complementary colors; they lay green eggs and have a brown head and a gray backside. Alycia loves her unique flock and equally colorful egg collection, but it means she ends up with a lot of roosters. Not only does she buy unsexed chicks but the reproduction that takes place in her yard produces them as well. At this point, although they have eaten them in the past, they generally prefer to rehome them. "The meat is so tough and it takes such a long time to raise them to eating age. It just didn't really seem like a viable, sustainable source of protein for us," she says. Last year, the rare-breed roosters that Alycia ended up with were so small that they were "not really worth a meal" in addition to being adorable. "I couldn't bring myself to do it." They found someone to adopt all of them, and "now they're a flock of weirdos" living on a farm in nearby Fairfield.

Alycia and Kevin have been working to create a sustainable, backyard agricultural system that can provide for their needs and give them an incredible degree of self-sufficiency. Alycia tells me that she wants to create a system in which nothing is wasted or left behind. Their yard is planted thoughtfully, the bees and plants and herbs

CHAPTER 3

and chickens and flowers and rabbits put together in a mosaic that makes the most of the space. On the surface it might seem like a lot of work, but when Alycia talks about connecting to the earth and her mom through gardening and knowing she's ingesting nutritious food, I'm not so sure—in a lot of ways, this system seems easier than the alternative.

"I hated going into Whole Foods," Alycia says, "and seeing how much they were charging for a bunch of kale, and it would be way cheaper and taste way better and it'd be in my backyard so I wouldn't have to go do the freaking store to buy it." I think about how sometimes the "easiest" thing is both simple and hard, because the other option is just more convenient but undeniably inferior. Handling all the ramifications of those choices—the moral reckoning, the environmental effects of almost any industrial system, the animal welfare—takes either more time to digest or a level of isolation and ignorance to withstand.

Giving in to systems you don't agree with, feeling alienated from your food, and paying for overpriced produce also seems like a lot of work. "We have a lot of personal and emotional reasons why we eat this way," Alycia says. "It's a way for us to be independent and give the middle finger to the poultry industry and our food systems that are so dependent on the disconnections that we've created between our food and ourselves."

One afternoon when I'm heading outside with a bowl of kitchen scraps for the chickens, dogs in tow, I wonder about the disconnect between people and food. As we head into the yard, Atlas and Tashi bound across the grass. Atlas, a small dog with fruit-bat characteristics (both in his giant ears and food preferences), makes a quick U-turn to beg for a strawberry top in the bowl I'm carrying. I grant his wish. He takes it, runs to a shadowy spot, and begins to munch. Tashi, totally uninterested in fruit, sniffs the grass, finds an appropriate spot, and rolls onto her back with her belly exposed to the sun.

While I feed Francis, Grendel, Joan, and Amelia the remaining strawberry tops and watermelon rinds, I imagine what it might be like to have a more closed system in my own backyard that is more independent and less reliant on outside sources.

I pick some of the early fall figs and share them with the chickens while Atlas tears in circles with the hide of a gutted stuffed animal in his jowls and Tashi, belly up, huffs and sighs and scratches her back on the concrete. Joan gets fig all over her jowls and wipes the sticky fruit off on the ground as she moves her beak and head back and forth in long, dragging motions. This is a common way that chickens clean their beak and face when they're done eating.

The dogs are restless, so I let them stay outside and play without me. When I come back ten minutes later to check on them, Atlas is standing over Francis's body. He looks proud and wags his tail as I approach. Tashi is lying on a rug on the patio. "Inside, inside!" I yell to the dogs before lifting Francis's head from the lawn. She is warm and soft and dead.

I run to the coop, looking for the others. Amelia is plunged halfway into a stump, not moving. Joan is on the topmost level of the coop, safely tucked away. I cannot find Grendel.

It's hard to rationalize what happened. Tashi had been around the chickens since they were only a few days old. She'd let them sit on her; they fell asleep against her side. While Atlas had always been a little too interested when they were face-to-face, he'd never broken into the coop, and when the neighbor's chickens fly over the fence

CHAPTER 3

he never gets too close. This is the dog who was once attacked by a duck—and lost. Tashi and Atlas have been outside almost daily since the chickens had been in the coop and nothing has ever happened before.

I start collecting the bodies as I search for Grendel. I start with Amelia. When I try to remove her from the stump she tenses, balks, and rears. "You're alive!" I scream, running my fingers over her body. There are no signs of blood and nothing is broken. I put her in the coop beside Joan.

Grendel is at the other end of the yard, also dead.

I dig a hole by the orange tree, drop in Francis and Grendel, cry, and apologize to their corpses. While I am casting layers of dirt over them I think about cleaning their vents, hand-feeding them, and those twin desires I once experienced—wanting them to be useful and wanting them to love me. I promised to protect them and didn't. I thought the run was secure and it wasn't.

This is one of the hard truths of urban agriculture. It's easy to forget that it's actual agriculture, not just play-farming with happy hens and a garden of cucumbers. If you want it, you work for it, and if I want to eat chicken, I need to raise hens. It's that simple, but this was not my intended purpose for Francis and Grendel.

Later that afternoon I begin to wonder, *Should I have eaten them?* The question seems crass. Perhaps it's perverse, but I wonder what is most right: burying them like they're pets or utilizing their bodies so as to not "waste" them? Do they have to be made explicitly purposeful? Do I even have the right to consume them since my human error resulted in their death?

I picture Francis in scalding water and plucking her fluffy breast free of gray feathers, about notching Grendel's knees with a sharp knife so the tendons sever and I can remove her feet. It doesn't feel right. I've spent weeks anthropomorphizing them, calling them names, posing them for pictures, and hand-feeding them. Will I treat the chickens I buy for meat the same?

No, and yes—clearly yes. I will not put them in a different coop, I will feed them the same food, I will check on them regularly. I will stroke them and spend time with them and provide for them just as much as I do for Amelia and Joan, as much as I did for Grendel and Francis. But I will also eat them, at least supposedly.

I spend the rest of the day compulsively checking on Amelia and Joan, holding out a small dish of water or a palmful of food for them, and coaxing them to eat, to relax, to be okay. By the afternoon, they are taking dust baths, scratching in the dirt, and voraciously consuming worms. After a couple of hours, Joan stops calling for the lost members of her flock.

In the coming weeks, I find that Joan and I are more alike than we know. Both of us do what we need to do to carry on—for her, this means resuming normal rhythms of life and forgetting that Grendel and Francis existed. She becomes a one-chicken type of hen and follows Amelia everywhere. For me, doing what I need to do means something else entirely. It becomes clear that caring for my chickens and continuing to move toward a more interactive, personal relationship with my food also means concerning myself with where Joan and Amelia's food comes from. While Joan follows Amelia, I end up following the trash.

4

A Freegan Flock

DUMPSTER DIVING AND LIMITING WASTE

"**I**S THIS ILLEGAL?" my friend Kari asks. We are on our way to a superstore dumpster wearing long pants and galoshes, our hair pulled back because we mean business. Headlamps are strapped across our foreheads like sweatbands, and we've stashed several reusable bags and a bottle of hand sanitizer in the backseat.

Dumpster diving feels like stealing but isn't. The waste our food system generates—trash that takes resources to produce and then dispose of—is astronomical. By wallowing in the glut, we've backed ourselves into a dire environmental corner, so even though I'm not the kind of person who readily buys fancy cheese, premade garlic bread, or individually packaged slices of chocolate cake, I will nab them from a dumpster.

"Technically no," I tell Kari as we pull into the parking lot, "but it's as good as if it were." In 1988 the Supreme Court ruled in *California v. Greenwood* that police didn't need search warrants to go through trash that had been put on a curb. In other words, when someone throws something out, it's in the public domain. According to the ruling, "It is common knowledge that plastic garbage bags left on or at the side of a public street are readily accessible to animals, children, scavengers, snoops, and other members of the public." Crawling into a dumpster is arguably just as legal as perusing a trash bag on the curb. The store doesn't own what it threw away, and the garbage collection company doesn't necessarily own it either. Dumpsters are public containers: bring on the scavengers and snoops. I'll happily haul reusable bags and headlamps to that party.

Dumpster diving seems weird at best, gross and unsafe at worst. But public awareness of dumpster diving, just like public interest in backyard chickens, seems to be on the rise. It's been joked about on *Portlandia*. There are countless how-to videos on YouTube, although anyone and everyone thinking about dumpster diving should proceed with caution and talk to experienced divers if possible. The *Guardian* published an article on how certain individuals living in New York City can achieve a middle-class life on less than $5,000 a year with the aid of local dumpsters.

In the dumpster Kari and I find several flats of frosted cupcakes, ten large tubs of baked beans, four bags of pita chips, and an entire case of frozen TV dinners. We pile lemon meringue pies and baguettes into the car's backseat, toss aside empty cardboard boxes, and uncover certain items we elect to leave behind—plastic packages of bacon, frozen seafood, and mushed fruit don't make the cut.

I start actively dumpster diving and reduce my grocery bills by almost 80 percent. Our kitchen remains full with sweet potatoes, onions, garlic, broccoli, apples, peppers, carrots, green onions, cucumbers, corn, celery, baked goods, bread, and even the occasional coconut. I make gallons of soup and sauce. One of my favorite stores begins stocking almond milk on their shelves, which means I get it

CHAPTER 4

from their dumpster. Although Katie is rightfully skeptical of my strange new habit, she admits that she's getting used to it, especially when I bring home such good stuff. We make sure to cook things well, and I'm constantly mindful of perishables, store closing times, and the proximity of edibles to the walls and floor of the dumpster, which are notoriously dirty surfaces.

Needless to say, dumpster diving is definitely not for everyone. My family worries that I will find something in a dumpster—a needle, an animal, hazardous waste—that will result in sickness. I'm even coached not to let a lid close on me lest I suffocate inside the metal belly of a waste receptacle. I want to say clearly and unequivocally that I cannot advocate dumpster diving for others, nor will a thin line about *California v. Greenwood* fend off descending police officers. I would say to never, under any circumstances, climb into a trash compactor, but always use common sense, be safe, know your limits, and bring a buddy. I'm not interested in sending anyone tumbling headfirst into the trash unless I'm holding a headlamp for them. People are sometimes taken from inside of dumpsters into garbage trucks, where they can sustain brutal injuries or die. It has happened before. For me, skimming off a small level of surplus, determined to be waste by its location (in a dumpster) rather than its quality as food, makes good sense, but this activity isn't without risks and shouldn't be undertaken by those who don't have, at the very least, an experienced friend to show them some safety protocol.

Collectively, Americans throw away 150,000 tons of food every day at the rate of a pound per person. This waste doesn't dematerialize. Discarded food, mostly fruits and vegetables, releases methane into the environment. Methane, like carbon dioxide, is a greenhouse gas, but unlike carbon dioxide, it is incredibly effective at absorbing heat. Although it doesn't linger in the atmosphere for long periods of time, methane is eighty-four times more potent than CO_2 and accounts for approximately 25 percent of our current phenomenon of climate change.

It's common knowledge among backyard chicken keepers that kitchen scraps can supplement their chickens' diet. Leftover rice and beans, bread crusts, strawberry tops, watermelon rinds with a small amount of fruit left—all of these are excellent chicken treats. The hens, in return, convert these unwanted or inedible foodstuffs into eggs. Not only do they turn calories we won't consume into ones we will, but they decrease the amount of food waste that ends up in the compost or dumpster.

When Joan and Amelia start laying eggs, I'm rich in animal proteins for the first time in a long time. Amelia churns out an egg almost every day while Joan manages four to five a week. The process of a chicken laying an egg seems innocuous, but it's actually somewhat complex.

Chickens only have one orifice, called their vent or cloaca, and this is where everything they expel from their body, eggs included, is released (however, the uterine tissue will move out with the egg, so the egg doesn't come in contact with the other unsavory substances that are also passed by this strange, multipurpose organ). Hens contain thousands of potential eggs at once. Each ovum can develop into a yolk, but this doesn't happen one at a time. Multiple yolks grow and develop in a chicken's body at once: if you were able to look inside a hen who actively laid every day, you would see a series of underdeveloped yolks ranging in size from pinheads and marbles up to their standard proportions.

The process of forming a shell, which involves calcification and mineralization, takes approximately twenty hours, although all prior steps, including the development of the egg white, take only about

CHAPTER 4

four. If a chicken doesn't have enough calcium in her diet or is overheated, her shells might become soft and flexible, much more like a bag than a brittle layer.

Around the time she starts laying eggs, Joan undergoes some behavioral changes. One night, Katie agrees to put the chickens in the coop for me. When she goes outside in the near dark, the chickens are already roosting. Joan leaps from the coop, charges Katie, and squats down in front of her. With her wings outstretched, legs bent, and head down, Joan's whole body shakes slightly.

"I thought she was going to attack," Katie told me after.

It makes sense. Joan is normally sweet and quiet. She's nonconfrontational, lets Amelia get the best grub, and hardly ever chatters, while Amelia will do a chicken version of a full-on scream if she sees you walking into the yard with a bowl of kitchen scraps. If Amelia is the chicken who yells, "Hurry up and feed me! What took you so long?" Joan is her patient counterpart who waits with her hands folded in her lap. To see such a passive, appreciative chicken throw herself out of the coop and charge you in the dusk before getting into a quivering sumo stance is unnerving. Katie, not knowing what to expect, ran back inside and waited until it was darker, and thus the chickens sleepier, to lock the coop.

When we talked about it later, I explained what the squat means.

"She was trying to mate with you," I tell Katie, laughing. The position Joan took is called a submissive squat, and it's a mating position. When a chicken squats down like Joan did for Katie, a rooster can hop onto its back. This process involves a strange act called a "cloacal kiss." Instead of penises, roosters have cloaca, just like hens do (roosters do have testicles, but these are internal organs, located along their back). In order to mate, the rooster will touch his cloaca to a hen's, transferring sperm from his body to hers.

Joan begins squatting almost obsessively. Without a rooster, she begins to see almost any human as a possible suitor. It's endearing, but also somewhat dangerous; when walking behind Joan, you have to be prepared to stop or dodge her when she goes suddenly stiff. This process isn't uncommon for chickens, and the submissive squat is often seen in a hen a few days before she starts laying eggs. Joan, however, takes it to the extreme.

After a particularly heavy eggplant haul, I begin making deep-fried eggplant sandwiches, using Joan and Amelia's contributions for mayonnaise and egg wash. One evening mid-fry, still contemplating the produce I left behind and how else to manage the waste besides sharing, bringing friends dumpster diving, freezing, and stockpiling, the answer hits me like a ton of slightly bruised apples.

I don't have *only* two mouths to feed: I have two mouths and two beaks. Joan and Amelia's feed is organic, sure, but the first three ingredients are corn, soybean meal, and wheat. The majority of organic corn production in the United States occurs in Wisconsin, Iowa, Minnesota, Michigan, and New York—all thousands of miles from where I live—and thus the ingredients have obviously been transported over roads and rails, from farm to mill to factory to distribution center to retail outlet where I toss it over my shoulder and head for the checkout line. Commercial corn consumes nearly 6.6 million tons of nitrogen annually (5.6 million tons come from chemical fertilizers, while almost a million tons are derived from animal manure), and all of those fertilizers can, and often do, contaminate groundwater, resulting in dead zones like the one near the mouth of the Mississippi River in the Gulf of Mexico.

CHAPTER 4

Soybeans and wheat are also problematic. Although soybeans can help fix nitrogen in soil, like wheat they're an annual crop. Annual crops, which must be sowed and replanted every year, require large amounts of labor, fossil fuels to run expensive equipment, and fertilizers to satiate the hungry soil. Monocultures—large swatches of just one crop, which is the conventional method in industrial agricultural systems to grow soybeans, corn, and wheat—deplete soils of their fertility, decimate their natural structures, and make them more vulnerable to erosion.

Annual crops offer little protection for wildlife among the rows, and while a plate of tofu doesn't necessarily contain meat, it is in a sense quite bloody. According to a 2018 paper published in the *Journal of Agriculture and Environmental Ethics*, "Depending on exactly how many mice and other field animals are killed by threshers, harvesters and other aspects of crop cultivation, traditional veganism could potentially be implicated in more animal deaths than a diet that contains free-range beef and other carefully chosen meats."

If purchasing or avoiding certain foods constitutes implicit support or condemnation of products—if voting with my dollar is possible—then this logic seems to go beyond my food and to the food I purchase for my chickens as well. I don't want to eat petroleum and mice, but that's what my chickens' eggs require if one were to track the long chain of inputs from the beginning: my chickens eat feed, which uses grain and corn harvested by petroleum-powered machines that kill small rodents in the field.

Dumpster diving for chicken and human food might seem radical, but the animals and I are not alone. Freeganism, a movement that advocates for removal from systems of production tied to negative capitalist structures of exploitation, has been on the rise for years. It combines the words *free* and *vegan* because while vegans avoid contributing to animal suffering, freegans avoid contributing to a capitalist system. Freegans limit their participation in the mainstream economy and focus on minimalism, waste reduction, and careful resource distribution. Under this ideology, participation in our economy is seen as a form of complacency that accepts things like sweatshop labor, deforestation, and factory farming.

Freeganism might seem strange, but it's been reported on by the *Guardian* and *Vice*. In "Not Buying It," *New York Times* reporter Steven Kurutz wrote, "Freegans are scavengers of the developed world, living off consumer waste in an effort to minimize their support of corporations and their impact on the planet, and to distance themselves from what they see as out-of-control consumerism." Kurutz continued to describe one individual who claimed, "'If a person chooses to live an ethical lifestyle it's not enough to be vegan, they need to absent themselves from capitalism.'"

I am definitely not ready for a step that radical, but I appreciate that dumpster diving can help me avoid supporting industries that still have a place in my life. I don't grow my own food, and it has to come from somewhere. I meet a dumpster diver, whom I'll call Nina. I don't know her real name because much of what we talk about focuses on her job—she's in the New York catering scene—and she often salvages food from work that would otherwise be thrown away. It's a practice that could land her in trouble, but she does it anyway. She sounds equal parts badass and environmentalist, and I can't shake the imagined image of Nina in a crowded New York subway car with bulky reusable bags on each shoulder and a tray of rescued food in her arms. In peak catering season, 80 percent of the food in her household—which feeds her and her partner—is salvaged.

"Some people ask why I bother dumpster diving," I said when we talked on the phone one morning, "because no matter what I'm not

CHAPTER 4

going to make a dent in food waste, and there's no way I'll ever salvage enough to make a real impact. What do you say to objections like that?"

"Refer them to Rebecca Solnit's essay on naïve cynicism," she said simply, before adding, "if they're the type of person who would take reading recommendations."

The essay to which she was referring, "The Habits of Highly Cynical People," examines a mindset of cynicism that, Solnit claims, "bleeds the sense of possibility and maybe the sense of responsibility out of people." The naïve part is expecting perfection and claiming to be "world-weary" when merely expressing an immature desire to collapse complex arguments into simple, straightforward binaries. Naïve cynicism responds to initiatives to address climate change with the assertion that we're so colossally FUBAR that bothering to change things is useless. Solnit writes, "Naïve cynicism loves itself more than the world; it defends itself in lieu of the world. I'm interested in the people who love the world more, and in what they have to tell us, which varies from day to day, subject to subject. Because what we do begins with what we believe we can do."

Nina told me, "There's this thought that everything is so bad that if you're really aware, you are angry and you are hopeless because you know how bad it is. I don't think that has to be true, and I think that's really counterproductive. We can all do something and it may not be enough, but it'll definitely be better than nothing." And even if something won't save the world, that doesn't make it not worth doing. Even if backyard chickens are a small form of resistance, that doesn't make this practice worthless. There's no one action that's going to save humanity and all of our natural spaces, but that doesn't mean it's impossible. It just means it's complicated.

A naïve cynic might say that our consumer society is so enmeshed with our daily lives that just by existing we're having negative impacts on the rest of the world. They might say that freeganism—real freeganism—is impossible and that if we really wanted to not commit evil we'd have to do what Darren Chapman of Tiger Mountain Foundation did when he was a young adult and move to the woods, forage for berries, and live in a cave. (He said that after seeing all the things he saw, he didn't want to engage with the world anymore.) Moving into the forest doesn't seem like a viable solution, but doing less evil does.

I know that my way of raising chickens in my backyard isn't a totally perfect system. They require pine shavings for their nest boxes; I've made the run out of new, rather than recycled or salvaged, materials; their commercial feed comes in non-recyclable bags; and their existence has exponentially increased the number of times I head to the local hardware store in search of materials to repair the coop, heat lamps, or feeders. Do I know where the pine shavings come from? Do I introduce myself to a fundamental truth about the world when I buy red bulbs for heat lamps? No. Of course not. It's not perfect, but the point is I try. Maybe it's impossible to do no evil and our best chance is to just do less evil.

The idea of a freegan flock isn't exactly in line with the mainstream backyard chicken movement, but it's far more similar than divergent. A capitalist system doesn't offer the choice of opting out. I know that my free eggplants are some of the spoils of capitalism, and I'm only able to get them because society's economic framework has decided that it's more profitable, more rational, to throw away good or slightly damaged food than it is to find a use for it. Sure, dumpster diving for chicken feed seems like a way to be more self-sufficient, but it's also a practice that's enabled by the system I want to disavow. At least, I tell myself, I won't be contributing with my money.

CHAPTER 4

The backyard chicken movement often seeks to implicitly or explicitly promote self-sufficiency. Chickens help create closed-loop systems of food production by providing fertilizer and eggs, then eating food scraps left over from the garden's spoils. Although feeding my chickens from dumpsters is an entirely uncommon take on self-sufficiency, I decide that it is worthwhile because it keeps a small amount of waste out of landfills. Even though this might be an insignificant drop in the ocean of edible food trash, I will not be a naïve cynic. I will not imagine that it doesn't matter.

Inside a dumpster with my feet carefully planted on the flattened cardboard of shipping boxes to better distribute the weight lest I sink into the unstable sea of produce below me, I bend at the waist and drive my arm into a small crack between the hard sides of two black plastic crates. Below, I can see a pile of semi-squished yet edible avocados. My reusable tote bag is slung across my chest like a sash, and I heave the green fruit into it as quickly as I can unearth it. There is a bag precariously situated above the black crates, but moving it would likely tear open the thin plastic where it is stuck to the side of another box.

In the process of dumpster diving for the birds, I've become a bit more cavalier. Shorts—with extra reusable bags tucked into my belt like some ultra-chic Recycling Faerie tutu—have replaced long pants, my previous dumpster uniform of choice. My arms are streaked with the wet pulp of the tomatoes I dig past. I am not bothered by either of these things.

I've found that dumpsters tend to have layers, and they don't dictate levels of decomposition or grossness as one might assume. This is the case today: although some of the tomatoes and bananas inside the bag above the avocado cave are smashed beyond recognition, the avocados are nearly perfect, and if I wasn't prioritizing Joan and Amelia's caloric needs tonight because I already have a freezer full of prepared dumpster food, I'd think about making guacamole when I get home. As it is, I plan to remove the pits and skins, which are toxic to chickens, and spoon the soft flesh into freezer bags so I can dole out the fruit over the course of the next week.

Chickens have a number of dietary requirements beyond their caloric needs. They require protein, essential amino acids, calcium, nitrogen, and vitamins A, D, and E. Avocados contain protein, essential amino acids, calcium, potassium, and vitamins A and E. They also have omega-6 and omega-3 fatty acids. I crouch and extend my entire arm into the shoulder-deep crevasse, keeping my head pulled back to avoid contact with a pile of sweet potatoes, to reach a few extra avocados before scrambling to another corner of the dumpster.

I lower the full tote bag onto the gravel parking lot from the open metal mouth of the dumpster and pull free a bag from my belt. I begin filling it with the sweet potatoes, which I plan to cook and mash for Joan and Amelia. It's possible to feed chickens raw sweet potatoes, provided you grate or chop them, but my chickens are discerning and picky; I've found they prefer theirs cooked. I'm particularly happy with the sweet potato stash swelling the tote bag. Sweet potatoes contain protein and are high in vitamin A in the form of beta-carotene, which is good for those bright orange yolks backyard chicken owners covet. Some estimates suggest that sweet potatoes can compose up to 30 percent of a chicken's overall diet.

As I bring my bags to their overflow capacity, I think about the produce I'm leaving behind and all the waste it represents in all the dumpsters across America. If we could take back the land and resources it uses to produce all the food that we turn into methane-belching trash, America would regain 30 million acres of land and kick a 780-million-pound pesticide habit.

Nina is particularly familiar with how corporations and businesses often find that the easiest way to deal with excess food is to pass it off as trash. Working with catering companies for the past three years has taught her about all the things capitalist companies throw away. "There's all this weird secrecy in the catering industry about how much waste there is. I think that's due to the fact that catering companies know that it's not a good look, both from a sustainability standpoint and from a product-value standpoint." Everything that goes on the buffet but remains uneaten goes in the trash, in addition to all the extra food that's never presented before guests. In the United States, according to the USDA, it's estimated that 30 to 40 percent of all food is wasted.

"The name of the game in catering is you have to have enough that there's way more than you need. They always say in a buffet situation that they want it to look abundant—huge piles of cheese on the buffet at all times, right up to the end—so you have to have an excess of what people are eating to create this effect." When meals are sit-down, served-to-the-table affairs, extras are still accounted for. "What if you have extra guests show up? What if something goes wrong and a sheet of it gets burned?" Even when guests are asked to select their entrees in advance, catering companies still bring more than anyone could eat. What if someone changes their mind? "They don't want to tell the guests no, so they have to have way extra."

Chris, another dumpster diver, also understands the amount of waste industries produce, although he's more familiar with grocery stores. Chris lives in Boston, but he started dumpster diving when he was traveling around Australia in a Volkswagen. One day he was heading to Queen Victoria Market, a large farmers' market in Melbourne, but didn't get there in time. They were closed. He saw a bunch of perfectly good food on the ground and loaded it into his van. Australia has a huge food waste problem, and his friends who lived there told him about dumpster diving at supermarkets. One afternoon when he was driving by a supermarket dumpster, the lid was open. A cereal box caught his eye. "From then on almost all of my food came from the dumpster."

On a recent dumpster run, Chris encountered his first "No Dumpster Diving" sign. "I've never seen that before. It said, . . . 'Violators will face fines,' or something like that." He found it upsetting: why was there a sign that banned dumpster diving above a huge haul of perfectly good food?

"It's so ass-backwards on what should be illegal. Like, why is it illegal for me to salvage food when it should be illegal for them to waste it?"

Freegans and dumpster divers have a lot in common, but there are some differences. Dumpster divers, like me, might still participate in the capitalist economy in real and pervasive ways (I drive a car, work for a state university, and buy toilet paper and beer and coffee and gasoline), and freegans might not necessarily dumpster dive. If freeganism and dumpster diving are theoretical siblings, the zero-waste movement is a not-at-all-distant cousin. While freeganism adopts an arguably more anti-capitalist stance than the zero-waste movement does by trying to opt out of the economic system altogether, the similarities between the two are startling. Being zero-waste means trying to avoid producing trash: no plastic bags, no bulky packaging, no wrappers. Metal straws and reusable water bottles might be the most visible hallmarks of the zero-waste movement, but it runs so much deeper than that. Consider the fact that

CHAPTER 4

Americans, on average, produce 4.4 pounds of trash per person per day, generating a collective 254 million tons.

Hilary Near, a commercial zero-waste assistant for the city and county of San Francisco, recalls visiting a landfill when she was in fourth or fifth grade and feeling like what she saw didn't make sense. Now Hilary helps businesses reduce their waste, aids with composting, and supports San Francisco in its goal to become more sustainable. "From a young age I just realized that the way I think our resources should be treated and the way they are are just not matching," Hilary tells me. "I'm hoping in the rest of my career . . . to not just be communicating about where you put things, but how to move on to a more comprehensive concept of how we use resources and just going back to that hierarchy: reducing before reusing before recycling." She works in youth education, replicating in some ways the experience of going to a landfill that shaped her, but she also works in policy, helps businesses begin composting, and encourages food recovery instead of food waste.

It's eleven at night and the dumpster gods have not yet been gracious. Wading through the dumpster outside of a chain coffee shop, I realize that the bags of baked goods that sent me over the metal lip all have grounds and milk poured into them.

"Are you serious?" I ask aloud in sheer exasperation before looking at my parked car. Tashi hangs her head out of the rolled-down window. "Tashi, are they serious?" I find two more similarly mistreated bags.

My other stops have also been disappointing. The pet store dumpster that usually has pristine packages of dog treats and bags of organic dog food was full of ripped-open bags of dry kibble, dirty cat litter, and what appeared to be used bedding from the small rodent cages, none of it bagged or separated. Leaning over the metal edge, I thought longingly of the time I found an intact kennel with the return receipt still taped to the side.

The dumpster that normally has ample produce was suspiciously empty. There were trash bags with paper towels, coffee cups, and cigarette butts, but gone were the squash and apples.

Pulling into the superstore parking lot, I don't have high hopes. "I'll be right back," I tell Tashi, who is acting as co-pilot from the passenger seat. "Paws crossed, big dog." There are two dumpsters: the first is empty. I close the sliding metal gate and move on to the next. There's a laminated sign that says, "Keep Closed At All Times" on the door. I grab the handle and push it open, the metal squealing against the frame.

The dumpster's interior is heaped with whole cooked chickens still in their closed containers, pastries in plastic circles, and a box with the label "100% Beef" on the side, full of trimmings, some slabs of fat the size of both my hands, and long lean strips of red meat. Nothing here smells rotten. I run back to the car, grab my reusable bags, and ask Tashi, "Is it a bad idea to feed chickens chicken?" She doesn't respond. As a kid our neighbors would give their chickens table scraps, and the hens seemed just as enthusiastic about the chicken I saw them occasionally receive as the salmon skin and chunks of hamburger.

It's a common misconception that chickens are vegetarian. They eat bugs, will attack small lizards and frogs, and don't turn their beaks up at meat when it's provided. They are omnivorous creatures. An article in *Smithsonian* commented that although chickens enjoy seeds and insects, they will also consume "larger prey like small mice and lizards."

CHAPTER 4

Back in the dumpster, I begin shoveling the cooked chickens into my canvas tote bags, not knowing yet if I intend to feed it to the dogs or to Joan and Amelia. My strange pragmatism chills me, but I brush the thoughts away like flies and keep lifting chicken out of the dumpster. *I'll decide later*, I tell myself, although I do end up feeding it to the dogs. I grab a few racks of ribs, also in plastic, and a few bags of breaded chicken tenders. I heave an entire garbage bag of baked goods, not bothering to sort through, then fill one of my sacks with several loaves of slightly crushed but still perfectly decent bread.

Having loaded several bags of food into my car, I contemplate the box of beef trimmings. There are sheets of fat, strips of meat, and round cuts of marbled flesh. I crawl back into the dumpster and poke a piece, testing it. It isn't slimy. I smell my allegedly contaminated finger—nothing—then put my whole head in the box. My headlamp illuminates the small fibers on the flesh.

It's hard to see animal products go to waste. Meat and dairy require so many resources to produce. "I have a weird thing about vegetarianism and veganism," Chris told me when we spoke, "where I'm like, if it comes from the dumpster, I'm eating it." When he salvages meat, he finds himself thinking about the animal and what it went through. "It was just raised, resources were wasted to feed it, to make it grow, then it was slaughtered, and now it's just going right into a dumpster. It's hurting the environment to raise this animal, then hurting it even more to throw it in the dumpster and have it rot."

Here's the thing about industrial meat: regardless of anyone's moral position about whether or not animals should be consumed, it's impossible to argue that animals should be raised, slaughtered, and then thrown away. Maybe we don't need to raise animals in factory farms—maybe we don't need to eat animals at all—but there is at least some sort of utility or goal applied to their suffering when they are eaten. Not everyone is convinced that gustatory pleasure, access to non-plant protein, and family traditions like turkey on Thanksgiving are more important than animal welfare, but the meat industry does produce things that people make use of. The chickens in the dumpster run counter to this narrative.

Chris's voice rings in my ears. "What was the point of this animal's life other than to live and be slaughtered and go right into the dumpster?"

Max Wechsler, operations manager of Urban Ore, a reuse retail and recycling ecopark based in Berkeley, California, is aware of the environmental ramifications our addiction to waste has. He says that issues of plastic use and the overabundance of waste affect everyone: "Everyone needs clean air and clean water and access to non-polluted terrain . . . whether you're a Republican or a Democrat or whatever; those are universal quality-of-life standards." But changing a system, even if it makes sense for everyone, is a massive undertaking. The systems of trash production are culturally and economically entrenched into the fabric of our society. "Big Oil. Big Plastic. Big Litter," Max lists. "These are really tremendous institutions of capitalism."

Although Max has been building a career in waste reduction since 2011, he's been working with resource reclamation for as long as he can remember. When Max was growing up he spent half of the time with his mom, who adhered more to what one might consider traditional American ideals of consumerism and urbanism, and the other half of his time with his dad in a junkyard just outside of Philadelphia. As a kid Max could differentiate between different kinds of metals, sort scrap, and spot something valuable on the side of the road. As Max's father's residence turned junkyard, the neighbors demanded that a fence be erected. The looks of the place—produced from a

CHAPTER 4

combination of thriftiness, hoarding, and obsession—offended their delicate sensibilities.

Max used to have an economical take on recycling. It wasn't necessarily something cool and trendy. Productive? Yes. Financially beneficial? Definitely. But ideologically meaningful? Not quite. It wasn't until later, when Max was working with Urban Ore, that he woke up to the social and environmental ramifications of salvage programs. "We have a team of salvagers at the Berkeley Refuse Transfer Station, the dump, the last stop before landfill," Max says. "It's where all the trucks dump their stuff onto a concrete tipping floor and it gets bulldozed into containers and then just dumped off to landfill." Their crew picks through and rescues "the good stuff," which equates to about three tons every day. It's enough to keep Urban Ore's Ecopark Store—a three-acre facility with a thirty-thousand-square-foot warehouse that runs on salvaged goods and donations—stocked with cabinets, hardware, sporting goods, lumber, books, toilets, appliances, clothes, toys, and just about everything in between. "I was in the trenches, and I saw the mountain of stuff pile up every day and disappear, but it's not disappearing, it's just going to somewhere else. There are things you can't unsee."

I can't unsee the chickens nor the beef, and I can't unfeel the frustration of finding purposely contaminated baked goods behind the coffee shop, and I can't unlearn the devastating effects of monoculture crops, and I can't unthink Chris's words. *What was the point of this animal's life?*

How are the strips of beef fundamentally different than the whole chickens? Yes, each individual chicken is representative of a life that would be completely for naught if I left the body in the dumpster, but the box of beef trimmings has to count for something even if it isn't an entire individual. Even if it's gross. "It is such bullshit that it has no reason for living at all," Chris said. "It didn't provide any nourishment for anyone. When I take it, I'm thinking I saved the animal's existence or something."

I unbend at the waist and pull my head out of the box of trimmings. I'd asked the dumpster gods for something to feed the animals, specifically the chickens, and this seems like it. Dumpster diving feels like a small mainstay against the deluge of unwanted goods, an important path forged with a daredevil attitude and closed-toed shoes. I wonder, would it be grosser to leave the mound of fat and meat to rot, or to take it home, cook it, and serve it to hungry chickens? The wet box is disgusting, but I have a plastic liner in my trunk, and I know this is the last stop before it goes to landfill. I decide that although this feels like an extreme case, I would rather be squeamish about wasting something than about hauling around slabs of meat for a good cause. I'd rather get woozy from lip service than get skeeved-out by an opportunity I, in some ways, asked for.

"Well, screw it," I say.

I begin to pull at the side of the box. Screw waste, screw being shy about getting dirty, screw paying money for things that enact environmental violence, screw capitalism, and screw letting anything besides my core decide what I'm going to do.

I drag it to the open door of the dumpster, climb out, and find that I cannot lift the box while standing on the ground. It's too heavy, too unyielding. There's a shopping cart nearby and I wheel it over to the open side of the dumpster. I climb out of the small sliding window, into the metal basket, lean back inside the dumpster, heave the box to the metal lip of the opening, then, with one hand on the box, and the other on the shopping cart, I lower one leg to the ground.

CHAPTER 4

From there I can finagle the box free. *Yes,* I think, *this is resistance.* When Max and I spoke, one of his statements struck me particularly hard. "That's what we're up against," he said, "a profit-driven economy and society that's dependent on perpetual growth. It's unsustainable by definition." In a lot of ways we are dependent on waste. *Well, fuck that.* I lower the box to the ground and climb the rest of the way out of the shopping cart. I take a deep breath, lift with my legs, and run-carry the box straight into the open trunk of my car.

On the drive home, I wonder about all the other waste that the meat industry represents. Presumably, the choice cuts of the animal were sold at the grocery store, with this fat and meat being left behind to rot. But what about the organs? What about all the blood? I think about how the production line goes way, way back and that there are so many parts of an animal that consumers don't think of when they think about eating meat. It seems wasteful. I roll down the windows and sing-scream along to a mixed CD in triumph and because I feel like it might be good to preemptively air out my car. Tashi sticks her head out in solidarity.

At home that night I freeze most of the meat but chop several cups. In the morning I cover the scraps with water, mix in a bag of salvaged frozen black-eyed peas, and let it cook. An hour later when everything is soft and there are globs of white fat bobbing to the surface, I take the pot off the heat, give it time to cool, and puree the contents with a few handfuls of reclaimed spinach. The result is a greenish paste the consistency of smooth oatmeal. I bring it to the chickens along with torn-up strips of meat from the pork ribs and some of their usual favorites: banana, a dollop of leftover mashed potatoes, and a few tablespoons of the dregs left behind from making oat milk. Amelia snatches a piece of pork and runs away with it. Joan digs into a banana slice. I leave the bowl and attend to some things in the yard. By the time I've returned, it is clear they have enjoyed their meal. What is left of the green mash is pocked with peck marks and the chickens are cleaning their beaks on the grass.

One night, when a cop pulls up on me inside a dumpster, it becomes clear that the freedom with which I can dumpster dive is a function of my privilege as a young white woman in society. It is dark and the police officer rolls up beside the dumpster slowly. There are boxes of produce on the concrete, I am wearing a headlamp, and my vehicle is the only car in the parking lot. He asks me what I am doing, as if it isn't completely obvious, and I say something about feeding animals that I raise, although I intend to consume most of this food myself. He asks a couple of questions about what they like to eat, and then seems ultimately satisfied. I assure him I won't make a mess and he tells me, "Stay safe."

I'm still in the dumpster. I wait a moment, since it feels like our encounter should be over even though it isn't. The police officer waits, still in his car, before adding, "I just had to stop and ask you what you were doing. You don't look like a dumpster diver."

I pause a moment. Clearly, with my headlamp and current position nestled between organic apples and leafy greens, standing with only my torso visible from the metal lip of a waste receptacle, I am a dumpster diver. There are cardboard fruit boxes lined up in a row at the base of the dumpster, and it is evident they weren't placed there by employees. A combination of grapes, bagged salad mixes, avocados, apples, squash, and other assorted produce sticks out from the top of each. This dumpster, in particular, is located in one of Lafayette's wealthy, predominantly white neighborhoods, in back of a health food store.

I take his comment to mean that I don't look threatening or poor or like I need to be here at 10:30 on a weeknight. Even the idea of

CHAPTER 4

telling whether or not someone is threatening or poor or in need based on physical cues is fraught. I am in a brightly colored shirt and clean shorts. Essentially, what I believe he means to say is that I don't look like someone he'd expect to be hungry, which would disqualify me from doing what I'm doing. Because I seem to be some interesting young lady with a slew of animals and not an empty refrigerator, this becomes okay in his eyes.

Dumpster diving is not safe for everyone in all contexts. If I didn't have the racial privilege that I do, I believe the cop would have been more critical of my presence, at the very least, or potentially seen what I was doing as stealing. I am read as nonthreatening because I am a young white woman. I can "get away" with things like this. Before leaving, the cop says that if a store manager or worker ever asks me to leave that I should, but as far as he is concerned, it is okay for me to be doing what I am doing.

After months of dumpster diving, I find that I cannot feed my chickens on a completely freegan diet. I worry they won't get enough essential nutrients and their pickiness sometimes means they won't eat what I provide, even though I uncover plenty of protein: beef, pork, beans, black-eyed peas. Amelia begins the habit of pecking me when I bring her something she doesn't want and seems to prefer her food cooked and puréed. In the end, I settle for both; I buy them chicken feed and offer it freely, but I also give them rescued food. Although it isn't perfect, it reduces my dependency on store-bought chicken feed by more than half. I expand their run so they have more space to forage for insects. They start laying eggs and maintain a high production rate.

I try not to take it too seriously that I never completely feed my chickens for free. I know that I could do more. I could raise maggots in a bucket in the backyard and find more fruit and nut trees around town. There are weeks I only salvage enough produce for myself but not the chickens. Other times, after overly ambitious hauls that leave me swimming in food neither the chickens nor I want to eat, I find myself frantically pawning off stacks of coffee cakes. No matter how much I take, there's always more, and there are often things I can't save. When I asked Hillary how she deals with the hard realities of waste, especially when reducing it feels like a Sisyphean project, she said, "Connect with the deep care that goes along with that. My concern and some of the fear is also rooted in a lot of love for this earth and these people, and that helps bring the joy and spaciousness back into it."

Even if it's not perfect, my dumpster practices still scratch the itch that started this project. They satisfy the hunch that things could be done better. When I spoke with Max, he told me that he feels right in his choice to live a low-waste lifestyle that focuses on recycling and environmentalism. "It feels good because it's just objectively good." He explained, "I will be on the right side of history. You know what I mean? We look back on the civil rights movement and they were on the right side of history, and our grandkids are going to be like, 'What the fuck were they doing with Styrofoam cups?'" While I might not have totally succeeded in weaning my chickens off the teat of capitalism, I am making progress. Even people who feed their chickens only commercial feed are making progress just by keeping them in their backyards. The smallest of steps toward reducing household waste are wholesome, productive, and worthwhile. Toward the end of our conversation, Max reminded me it's not all-or-nothing. "Of course you get overwhelmed when you really see the scope of what we are up against, and it's easy to get depressed about it too,"

CHAPTER 4

he said. "But it's just like you do the best you can. . . . That's all people can do."

I agree with Max. Although the chicken movement might not be here yet, I sympathize with this end of the consumerism spectrum.

5

Pampered Poultry

DESIGNER CHICKENS AND THE PEOPLE WHO LOVE THEM

WHEN SEAN WARNER and Patrick Pittaluga, cousins and cofounders of Grubbly Farms, began looking into possible markets for their dehydrated bugs, they stumbled on the backyard chicken movement. Grubbly Farms is a US-based company that raises, harvests, and sells black soldier fly larvae as chicken treats. When the company was first getting started, they weren't settled on a target market. "At that point my dad had chickens," Patrick told me, "and he was telling me people in the backyard chicken movement were buying mealworms for their pet chickens, and that these pet chickens were treated just like you imagine a conventional pet to be treated: treats, toys, supplements. If it's something a dog has, there's an equivalent product directed toward backyard chickens."

Patrick is right. It might seem crazy that some people treat their chickens like others treat their dogs and cats, but only to non-chicken people. Not only are there chicken treats and gourmet chicken food

CHAPTER 5

but there are also chicken sweaters, chicken leashes, chicken harnesses, chicken car seats, videos on chicken training, and books about communicating with your chickens. It became clear that Grubbly Farms, which had modest beginnings in a laundry room, could capitalize on the growing movement's underrepresented market. Patrick and Sean wondered about going into human consumption, but, according to Patrick, "ultimately decided that Western culture was not ready for insects, particularly fly larvae, as a source of food." Instead, they found that farmers, especially those working with chickens and fish, were interested in the sustainable, natural source of protein that soldier fly larvae could provide. The problem was that even small farms required immense amounts of protein per week to keep up with their industrial demands.

Then the light bulb moment with Patrick's father hit. "The treat market was underserved," Patrick said. Companies that sourced their insects from China dominated the American market. Most of the dried bugs available for chicken owners were mealworms, which are lower in calcium than the black soldier fly larvae that Grubbly Farms sells. "Our product, being so high in calcium and protein, can assist the chickens if they're going through molting or if their shells aren't thick enough." Although some people treat their chickens to opulent lifestyles, there's still the underlying belief among a vast majority of chicken owners that their hens should produce healthy, high-quality eggs.

"Talking about treating them like livestock or eating them is hugely a big no-no," Patrick told me. It's true. Most backyard chickens are pets first, productive livestock second, and dinner never. It's pretty common for chicken owners to raise hens and eat meat but never consider slaughtering them.

According to a study published in *Poultry Science*, only 12 percent of respondents butchered their chickens in 2013 for meat (comparatively, 8.7 percent dispatched sick or injured chickens and 8.4 percent axed male chicks or roosters). Only 2.5 percent culled hens who had stopped laying or had become unproductive. Perhaps not surprisingly, rural residents were more likely to kill their chickens than urban or suburban residents. In some ways, this illuminates the scope and the ways in which chickens are counterculture: for those living in rural areas, it's more normal to slaughter your hens for meat, and for those living in suburban areas it's more normal to keep your chickens as pets. This divide seems to press the issue of whether or not chickens are a form of micro-resistance against a society that, for the most part, asks people to take a backseat to the production of their food. Can a pet chicken fight back against a separation between individuals and their food?

Probably. For those whose chickens are solely pets, they're still challenging what, or who, is worthy of human love. It's well accepted that dogs and cats are companion animals and not food, so by even having chickens it seems like some people challenge the appropriateness of whether or not they should be in industrial farming systems.

"It comes down to the mentality of looking at them as pets," says Patrick. The trends that impact the way people view dogs and cats also sway chicken owners. Both markets—the pets-as-pets market and the chickens-as-pets market—are subject to some of the same cultural and social influences, "similar to how, in the past ten to fifteen years, in the dog industry, people are becoming more concerned about what they feed their animals. There are brands now

CHAPTER 5

that either do human-quality or grain-free dog food," says Patrick. "We're starting to see that in the backyard chicken world: people want to do soy free, corn free, organic, no by-products and are starting to spend more money to ensure they get the highest quality products for their chickens." When chickens are pets, it implicitly justifies the idea that they should be given fancy, wholesome feed. However, it's not always so straightforward; unlike dogs and cats, chickens are expected to work for their room and board.

In a 2014 study published in *Poultry Science*, researchers at the University of California, Davis, found that the five most common backyard chicken breeds are Rhode Island reds, Plymouth rocks, Ameraucanas, Orphingtons, and Wyandottes: all chickens who are better at laying eggs than they are at converting calories to cutlets. Each one of these breeds lays colored eggs (respondents also expressed preferences for Easter eggers, chickens that lay blue-green eggs, and Marans and Welsummers, breeds that lay eggs the color of hot chocolate). Although it's possible to look at chickens as pets, it's improbable that they would've risen to this position if they didn't offer something in return, namely eggs.

According to Patrick, "People are starting to spend more money to ensure they get the highest quality products for their chickens and to improve the benefits of the eggs for themselves." He continued, "A lot of people, even though they view their chickens as pets, see the egg part as a very important factor."

Giving chickens dried bugs and treats isn't the only way to improve egg quality. One of the most effective ways to pamper chickens and help them lay the healthiest eggs possible is to give them access to fresh pasture. Allowing chickens to forage for grasses and insects improves their immune systems and levels of vitamins A and D and omega-3 fatty acids. It provides them with important sources of protein, fiber, calories, and carotenoids, which are vital for metabolic functions.

Chickens have been raised on pasture for hundreds of years and historically were allowed to roam and scavenge for most of their caloric intake. During the mid-twentieth century, giving chickens on egg farms access to greens could prevent disease and ensure flocks received the vitamins and minerals they needed. Farmers would often let chickens roam around the coop where they laid their eggs to ensure they were given access to spaces where they could forage, while other farmers grew and harvested specific crops—like barley, ryegrass, clover, and kale—to give to their laying hens. According to Jeff Mattocks, a livestock nutritionist, "Pastured poultry eat 5 to 20 percent [of their diet] from pasture, depending on the type and age of the poultry, and the quality of forage growth."

Some anti-pasture advocates argue that since chickens don't have a rumen (the organ in cows that helps them convert grass to food), they don't benefit from eating grass. Research, however, suggests otherwise. A 2019 study published in *Frontiers in Veterinary Science* shows that when single-stomach animals, like chickens, graze on pasture, the fatty acids they absorb are consumed and then transferred into their tissue lipids. This means that the chickens are healthier and their meat is too. Certain grasses, like pre-bud alfalfa, contain upward of 20 percent crude protein. When chickens are given free access to pasture, their omega-3 fatty acid levels rise. Additionally, the fiber in grasses and forage does two important things for a chicken's overall health. First, the indigestible fiber, which is fiber that isn't broken down by an animal's body as it passes through its system, helps give chickens more time to digest the rest of their food. It slows down their system enough to help them get the most out of what their bodies can break down. Second, digestible fiber feeds beneficial

CHAPTER 5

bacteria in a chicken's digestive system. By releasing lactic acid, digestible fiber effectively and positively stimulates gut health.

Chickens that can feed on pasture and forage for plants and insects are almost certainly happier than chickens who cannot; not only are they consuming the nutrients they need but they are also doing what they are made to do. Healthy eggs and pampered chickens go hand in hand, and giving hens the freedom to safely roam arguably improves their quality of life. Those dark orange yolks that are so coveted in the backyard chicken community as indicative of higher nutrient levels can be achieved by foraging.

That isn't to say dried soldier fly larvae isn't an incredible chicken treat. In fact, it's possible chickens enjoy it so much because they are creatures driven to consume other smaller creatures. Chickens that are allowed to hunt for small creatures and insects like worms, crickets, frogs, and slugs have higher levels of methionine, an important amino acid that is found in unfortunately low levels in grain. Without this vital nutrient, chickens will peck and consume feathers or even commit cannibalism. Black soldier fly larvae has high levels of methionine.

"Grubblies" give backyard chicken owners a chance to get close with their chickens. Patrick says, "The way people utilize it [Grubblies] gives them an opportunity to interact more with their chickens and feed them straight out of their hand." These things make chickens tamer, more accustomed to people, and more affectionate, but the bonding goes both ways.

Dave never expected he'd become so close to his Rhode Island red, Sammi. Dave, who has short black hair, glasses, and expressive facial features, lives with Sammi in Florida. He has a background in agriculture and worked as a high school teacher for years in eastern Colorado before meeting Sammi. "I taught poultry to my students among other things." Although Dave says he had good experiences with chickens, he looked at chickens as productive animals, like cattle. When he kept chickens, his approach was pretty traditional: the coop was in the yard, and the chickens were penned up in the coop at night. "That was the end of the engagement with the chickens."

Fast-forward to today. Some of Dave's fondest memories with Sammi involve bringing her on a train ("That was a very cool experience for her," Dave told me); being invited to a concert in Charleston, South Carolina, and being sung to by the band backstage; and going snowboarding together in Breckenridge, Colorado. They travel across the United States, swim at beaches, and go paddle boarding—together. "She has so much faith and trust in me. When she does get into an environment she's not familiar with, that's when I notice even more how much she finds security in me."

Dave didn't plan on having a close relationship with a chicken when he got Sammi; that's just how it went. In the wake of another pet's death, Dave found himself wanting a companion. He wasn't a cat person and didn't feel ready for another dog. "I don't know what my expectation was. It was quiet in the house and there was that internal feeling that we have to be responsible for something, to take care of something."

It didn't matter that Sammi was a chicken, for Dave wasn't necessarily interested in an intensely emotional relationship. "The thought to create a bond and express emotion or love for her and experience her expressing that affection toward me was so foreign. I never thought a chicken had that mental capacity."

At first Dave left her in a pen inside when he went to work (he didn't want to leave her in the backyard where predators might get her during his absence). Around the time she began laying eggs, Sammi started making a mess of her enclosure when she was alone.

CHAPTER 5

She would spill the water and the feed, lay an egg in the slop, and the mixture would combine with her feces. It was clear that something needed to change. "And I don't know if it was because she started reaching maturity and laying eggs that she started getting messier, but at that time I thought I needed to come up with something different." Dave began bringing her to work with him.

They started spending a lot more time together during his commutes. Dave noticed that when he checked on Sammi over the course of the day, she seemed more excited to see him than other people. "I was probably fighting that bond where she was craving that bond," Dave admits. "I didn't see it. I didn't know."

It took a while, but Dave got there. "I definitely have come to learn that that tiny brain—sorry—" he quickly adds, turning to face Sammi standing at his feet, "that tiny brain in there is capable of so much more than I have ever in my life given credence to." While chicken brains aren't necessarily small, Dave just didn't expect Sammi to express strong emotions toward him. "A farm animal can show and express just as much love, just as much companionship, just as much affection as a dog or cat," he says. While she has yet to lick Dave's face, she does show an incredible amount of intelligence in navigating social situations with Dave.

Once when Dave was driving down the interstate with Sammi in his truck, he scolded her for something. "I don't remember what it was," he says, but he remembers the way she reacted. When Sammi is in trouble, she often responds by getting closer to Dave, as if to smooth the relationship with physicality. "This particular time she ran over to my lap." A few seconds later, she turned around and tucked her head under his arm to snuggle against his ribs. "She sat there for a little bit, and then, I think she was like, 'I don't think I'm working hard enough to get back in Dave's good graces,' and so she stood up, put her head up high and—no joke—she pressed her head, the side of her face, against my mouth, almost like, 'Will you give me a kiss? Are we good now?'"

Instances like this have proved to Dave that chickens are capable of expressing affection. Although he believes people wouldn't understand this until they've experienced it firsthand, Sammi's actions, her attentiveness to social cues, and her seemingly innate desire to show affection illustrate that chickens can convey complex emotions and develop strong bonds to humans. "I think we have forever put chickens into a category and have not allowed ourselves to consider having a chicken as a companion." When people see Dave walking down the street with Sammi on a leash, they are baffled and seem to wonder, "How is this possible?"

Somehow, it is entirely possible. Sammi and Dave have almost become poster children of chicken companionship. When they go to the beach, Sammi and Dave like to take walks and swim if the conditions are right. Dave reminds me of that old saying, "madder than a wet hen," although it seems like Sammi's never heard it. "I refuse to go out in the water if there's a lot of waves," Dave tells me. When the water is undulating, it's hard for Sammi to see over the crests, even if they're small. "I can totally sense that it's a bit intimidating for her." Fortunately, on the gulf the water is often glassy. "When it's calm enough that she can see around, that's when she settles down and is totally relaxed and comes when I call her and we swim around."

After they swim, Dave and Sammi like to relax. Dave sits in a beach chair and Sammi sits in his lap. "The sun will come down on her and she will just fall asleep in my lap. She definitely enjoys that. Maybe it's the sound of the waves and the warm sun, but we don't do that very much because we get a lot of visitors." Dave and Sammi have become a pretty popular duo. When I ask him to explain, Dave

CHAPTER 5

continues, "I don't spend very much time just the two of us, just being quiet, because about every minute somebody comes up." About 40 percent of the people who come up to meet Sammi recognize her from social media, like Instagram (@sammichicken), and about 60 percent come up because of the "oddity" factor. Not that it bothers either of them—Sammi loves the attention and Dave loves advocating for her.

Dave says he doesn't think there's anything particularly special about his relationship with Sammi. It's obvious that Dave cares for and has a close relationship with Sammi (he ropes her into the conversation and speaks of her with adoration while she sits across the room in a chair watching him as we talk), but he doesn't think they're necessarily as unique as someone might imagine. When Dave got Sammi he hadn't heard of chickens as pets, but it didn't take long for him to realize that it had been done before. "I met so many people who said, 'I had a pet when I was a kid and she followed me everywhere and slept in the bed with me.' It was like, this is a real thing."

Although my chickens don't have it that good, they aren't slumming it. One afternoon in November, I explain to my dad that taking chickens on road trips isn't a crazy idea, although I honestly believe it might be.

When I tell Dad the Christmas travel plans, he asks, "Are you serious?" Even over twenty-three hundred miles and a spotty phone connection—he's riding his motorcycle somewhere on a coastal ridge in California—I can hear his understandable dismay. He exhales.

"When I come home for the holidays, everyone gets to go." I laugh, trying to make this seem funny, although I'm not sure it actually is. I'm in the backyard, kneeling in the chicken run. I turn over a large rock and Joan and Amelia charge. I have just proposed that we drive cross-country over the course of three days with two dogs and two chickens. "I can't just *leave* them here for a month."

"What about your neighbor?" Dad asks.

"You mean my neighbor who lets her chickens run amok in the street? No way."

"What about your friends?" Dad asks.

"Most of them are going home," I counter, which is true. I've somehow convinced myself that this is absolutely, 100 percent, without-a-doubt the most logical course of action.

"Okay," Dad responds. He knows how important my chickens are to me. My parents have a flock of their own, so the precedent of chickens in their yards has been set. Their Marin County home, just north of San Francisco, has an L-shaped yard, the foot of which contains a chicken run, a large coop inside of a secondary completely protected area, a small storage shed for hay and feed, a redwood tree that provides shade and shelter, and even a chicken swing (not that any of the hens use it). Besides one Rhode Island red, a "dual purpose" bird that lays eggs and converts a substantial part of its calories to flesh, making it a homestead favorite (not that they'd ever eat Big Red), all of my parents' chickens are fancy. There's a cream legbar named Minty, a silver-gray Dorking named Raquel, and a splash Cochin named Han Solo.

"Do you think they'd get along with our chickens?" Dad asks. In the background I can hear wind ripping over the ridge. It sounds like newspaper scrunching. Amelia snatches a june bug larva with her sharp beak, throws it back on the ground, then scoops up its thick body to run away from Joan so she won't have to share. Joan gives chase and Amelia turns to block her.

"Oh yeah," I say. "Definitely."

CHAPTER 5

When Willow was a few days old, her mother flew her via private plane from her birth state of Massachusetts to her new home state of New York. By all accounts, Willow has since grown into a city diva. She loves fast cars, pedicures, Sunday brunch with her sisters, and the Hamptons. Willow's mom works in an air hangar, and when Willow gets to tag along she entertains herself by watching the jets and helicopters take off and land on the runway.

At home, Willow loves to model. She tilts her head during photo shoots to give the camera different angles and poses beside Louis Vuitton boxes and Corvettes. Willow loves sweet potatoes, bananas, strawberries, corn, and mealworms. Her two feet are feathered, wings pure white, and she is incredibly fluffy.

Kristen was in her early twenties and working as a New York designer when she started contemplating chickens. As soon as she had her own place with a suitable backyard, she began working out the logistics of adoption. "Having a niche for quirky, odd-looking things, I fell in love with silkie chickens when I saw them on my Google searches." It's easy to see why.

Part cotton ball, part Muppet, silkies' feather barbs refuse to lock, resulting in a fluffy mass of wayward spires. They are relatively small chickens—roosters can weigh up to four pounds, hens only three—which gives them a baby-bird vibe, and their incredible head plumes blossom into spherical halos that defy gravity and put to shame any Midtown pompadour, although these dramatic hair styles often obscure their vision. Silkies' beaks look like something just brave enough to poke out from a blanket but not tough enough to emerge completely, and their legs and feet grow feathers, giving them a fashion-forward fur-lined boot look. In the thirteenth century Marco Polo chronicled the Chinese chickens with "wool" instead of feathers that he met on his travels (some experts believe he was writing what might have been the first ode to silkies), and in 1645 Italian naturalist Ulisse Aldrovandi compared their hair to a cat's.

Although they are adorable, silkies are arguably impractical. If you'd like to raise backyard hens for food, silkies are a poor option. They are unreliable egg producers, too small for a decent meal, particularly susceptible to Marek's disease and mites, and need to be toweled off or blow-dried if they get wet (their downy plumage lacks any waterproofing: a drenched silkie looks just as sad as a soaked tabby). Either because of these traits or in spite of them, silkies are one of the most popular ornamental chicken breeds today.

These small cartoon characters who run like drunk dinosaurs and are infinitely more hilarious than useful are the epitome of designer chickens. They make great pets but arguably crappy livestock.

Although I do not need more beaks to feed with trips to the dumpster, I seriously consider adopting a silkie or two just for the comedic pleasure of watching them. My back-to-the-land urban farming impulse has its limits. My practicality is finite. Somehow, I resist the urge to adopt my very own feather-footed Muppet but have no idea how.

After extensive research and literally sleepless nights imagining her perfect flock, Kristen found a US-based silkie breeder. She borrowed a company plane and brought her new chicks home in a shoebox. Three years later, she has six silkies: Aspen Louis, Poppy Picasso, Fawke Nova, Cash Monopoly, Turbo Polykarp, and—of course—Willow Mona. Aspen acts like a miniature Willow, Cash attempts to establish authority over the others although Turbo is the true leader of the group, Fawke is chick-crazy and wants to hatch

CHAPTER 5

more babies, and Poppy is, simply put, "special." According to Kristen, Poppy "often gets stuck spinning in circles." Willow gets special privileges because she's the quietest and best behaved, and Kristen brings her almost everywhere she can.

Kristen's chickens live a lavish life. They ride on fancy jets and get pedicures in the backyard. "Yes, they depend on us for food, shelter, and protection from predators," she says. "It's not necessary to go to the extent that I do, but I have so much fun with it." She links two of her passions—chickens and photography—on her Instagram account. "Willow and all my other chickens have been handheld since they were babies. I got them when they were just one week old. They've been modeling since they were babies, so they're not camera shy." Her rooster, Turbo, "poses like a king" when she photographs him. He's a "show-off" who loves attention, but he isn't too self-absorbed to be a responsible rooster to the ladies in his flock. "Turbo, if he finds a worm, will give it to the ladies and cluck really loud."

While Turbo enjoys walking on a leash, Willow prefers to be held. "My outings with Willow are normally to the office. I've always worked in aviation/aerospace hangars so there's always a little pen for my chickens at each one and she will stay there while I work," Kristen says. Willow's charisma is contagious. Everyone in Kristen's office loves her: "Some people even bring her veggies, lucky bird." When she isn't acting as secretarial support staff, Willow likes to visit the beach. "There's an incredible beach on Dune Road in West Hampton that she likes to go to." During the drive, Willow sits in a dog car seat or sometimes a small box.

PAMPERED POULTRY

With two dogs and two chickens, I pull up to the arrivals terminal at the Houston airport. I roll the windows down and start yelling for my father. The dogs, either in surprise or support, bark. He's riding with us back to California, convinced I might not make it without him. This is, after all, the first time I'll travel cross-country with so many animals. Joan and Amelia ride in a dog kennel I rescued from a pet store dumpster while the dogs vie for who gets to sit on the passenger's lap.

We sneak Joan and Amelia into hotel rooms by requesting a location near a door to the ground floor, then placing a sheet over the dog kennel with the chickens inside and carrying it in quickly. In the hotel rooms, I scatter mealworms in the bathtub for them to scratch and peck, let them wander the bathroom (cleaning up carefully after them), and give them food and water. One night there's an incident when I place Joan temporarily on the bathroom counter and she sees her reflection. She begins making loud *bruuup brrrrrruuup brrruuup* noises (which I can only assume the neighbors heard through the thin hotel walls), and then attacks the mirror. She flaps her wings to get airborne, and her claws make contact with the mirror. I have to snag Joan as she's mid-leap, and when she's placed on the floor she wanders over to Amelia as if she's just won an arm-wrestling match.

I buy the chickens bananas at gas stations and feed them through the bars of the dog kennel while Dad drives. I change their bedding at rest stops and mix their food with water to help them stay hydrated. Both Joan and Amelia lay eggs throughout the trip, which I take as a very good sign.

My dad might be the kindest, most accommodating person I know. He has built a temporary coop in the back corner of his

CHAPTER 5

chicken run. I haven't seen it in person yet, but I have seen pictures. It's elevated, with a ladder leading into a double-story plywood cavity. There's a second loft level with a roosting bar and two nest boxes, and the entire coop is in enclosed in its own wire run so Joan and Amelia can meet their aunts through the fence until they're ready to mingle.

From the photos Mom sent, the coop is well built, sturdy, and pragmatic. It is not painted or aesthetically paired with the house itself. There are no automatic doors, lights, or internal watering systems. It isn't what someone would call a luxury coop; it's an exercise in practicality.

Luxury coops aren't just about aesthetics, says Matt DuBoise, president of and designer at Carolina Coops. Matt has gray hair, strong-looking arms, blue eyes, and a warm smile. He talks quickly and passionately and strikes me as both practical and freethinking. On the topic of the interconnectedness between media, food production, and American's consumptive habits, he tells me, "You control the media, you control the food, you control the people," and I believe him. The people who control the media convince others of what to consume. Matt places emphasis on personal responsibility and is a proudly transparent business owner.

Carolina Coops is a family-owned and -operated business located outside of Raleigh, North Carolina, and Rochester, New York. Their most popular chicken coop design, their namesake "Carolina Coop," starts at $4,995 for a four-by-six-foot henhouse with one egg hutch and a twelve-foot run and goes up to $16,995 for a six-by-ten-foot henhouse with two egg hutches and a thirty-foot run. With the various add-on options, such as automatic run doors, heated watering systems with rain barrels and four-nipple waterbars, additional egg hutches, and cupolas, it's entirely possible to run up a bill of $7,255 or more on the smallest Carolina Coop model—and that's without a $1,500 custom paint job, $995 polycarbonate run cover, or board-and-batten siding. A large coop with ample accessories and a custom paint job costs more than the average down payment on a house.

For some it might be impossible to crunch the numbers without wondering who would fork over that kind of cash for a chicken coop, or why. But the price tag has utility and purpose behind it, and the reasons people pay it are clear and numerous. "I always tell people we have the best customers, but we have no two same customers," Matt says. "There is no specific kind of person who buys our coops. We have doctors, lawyers, and people who are blue-collar. We have a range of customers. Affluence doesn't correlate to what coop customers buy. It all comes down to what they see value in and what coop best fits their needs." Matt says there's no particular Carolina Coop demographic, although I know I'm not within it. As someone who lives off a graduate-student stipend, feeds her chickens from the dumpster, and knows that when the time comes to build a brooder for the meat birds the wood will be a salvage job, I suspect my chicken-related projects will continue to use pallets and plywood that have been abandoned on the side of the road.

There is one underlying feature that links most of Carolina Coops' customers: they see the value in investing in a solid coop that's been built to last. Carolina Coops cost what they do because they're worth it: they use metal roofing, which keeps the inside of the coops cool by reflecting heat; the windows are handmade in their own shop, unless otherwise specified, and framed with marine-grade wood; screws feature anti-corrosion layers; and the company maintains a practice of using pocket hole joinery and wood glue on the entire frame—a process by which holes, usually at angles of fifteen degrees, are drilled into one surface and then joined to a second surface with self-tapping screws. "Most of our customers are brand-new to chickens," Matt says, "but they've done their homework, and they say—and I love this—'We keep coming back to your website as we learn about the needs of chickens when it comes to coops, and now we see the value in your coops.' If you find out what chickens need, you're going to come back to our coops." In addition to carefully chosen materials, Carolina Coops has certain design practices that provide for chickens' fundamental needs, like giving chickens plenty of headroom above the roosting bar to more accurately mirror their natural sleeping habits (chickens roosting in trees have all that sky above them) and keep them cool (because the hot air from their breath and bodies rises). "When they learn about chickens, they realize why we do what we do with our coops and why we build them this way," Matt says.

While Matt's coops have always been beautiful, they haven't always been expensive. "In the beginning . . . when I had to build out of survival, I took what I could get and made the best of it." In 2008, during the recession and after recently moving to North Carolina, Matt headed into a dumpster to see what he could salvage. He needed materials to build things to sell. "My first coop came about that way, and that's how I learned to love my life."

As Matt climbed in the dumpster, he burned his hands on the metal. It was like an oven. Instead of letting go of the hot metal, "I grabbed on tighter." To Matt, the experience was elating. He could, with a little bit of bravery and some hard work, take trash and turn it into a valuable resource. It was like a treasure hunt. "When you find something in one of the lowest, worst places that people don't want to go, that makes you a better person," he says. "It's like Indiana Jones. He finds treasure in one of the deepest, darkest areas." Building nothing out of something made Matt realize how much better of a person he could be. It made him feel blissed-out. His first coop, built from salvaged materials, sold for $2,000.

But that level of work and fearlessness—stepping into a dumpster to haul out unwanted lumber and hopefully make something meaningful with it—doesn't mean things don't change. Now Matt buys high-quality materials from a store. "Before, I felt more like a badass when I had to take stuff that wasn't meant to be a chicken coop and turn it into one and generate business to feed my family. That's the true love in life, which redoes your DNA and makes you profound and strong." Matt argues implicitly through his experiences that venturing into the uncomfortable, using what you find and adding some labor, creativity, and a lot of determination makes people better. There's something about self-sufficiency that's supportive of autonomy. Even Matt's high-end customers, although they don't look in dumpsters, seem to be searching for self-sufficiency.

It makes Matt wonder, when he has a client who buys an expensive coop, if they know something he doesn't. It seems like they're "getting ready for something" when part of their retirement plan includes a small backyard flock. "They're not just building their dream home; they're doing ducks, chickens, bees, cows," he says. Perhaps, even though his customers can clearly afford luxury chicken accommodations, there is a more practical rationale behind this seeming splurge.

Once the six of us arrive in California in early December, my chickens take up residence in my dad's temporary coop, built from materials he had on hand in the yard and left over at work. It isn't fancy, but it retains the possibility for longevity; on a few occasions my dad,

CHAPTER 5

preparing to retire, says that he might keep the coop around so he can use it to acclimate new chickens before adding them to the flock. He, like so many chicken owners, wants to move beyond store-bought eggs and toward a more self-sufficient lifestyle of eating from the coop and from my mother's incredibly productive garden. After getting out from under his nine-to-five, my dad wants to spend more time collecting eggs, canning pesto and tomato sauce, making jam, and dehydrating figs.

Joan and Amelia quickly find the roosting bar, begin taking dust baths, and become accustomed to a steady stream of scraps (pasta, mashed potatoes, crusts) that come from my parents' kitchen. My parents' chickens spend a few days squawking and posturing at Joan and Amelia through the wire. The worst offenders are Han Solo and Big Red. Eventually, I release Joan and Amelia to join their aunts.

Skittish, my chickens wander the perimeter of the larger run. I take a handful of dried mealworms and goad everyone into close proximity with one another. Within a few hours when I return to feed them leftover pasta, they're all dust bathing together.

Jenny never imagined she'd be a crazy chicken lady, but the indoctrination process didn't take long. As far as I can tell, it's just something that happens to the best of us. You don't always see it coming.

"I would never go back to a life without animals," Jenny tells me. She lives in New York and has brown eyes and long brown hair. After starting a managerial position at Tractor Supply Company, she learned to care for chicks during one of their Spring Chick Days sales. On the last week of Chick Days, right after the last shipment,

Jenny knew she wouldn't be able to resist. "I loaded my cart and brought them home," she says. "It's completely different when they are yours." The bonding happened almost instantly. "Saying I was hooked was an understatement."

For a while Jenny and her husband kept chickens the "normal" way—outside of their New York home—but when a friend placed Eloise, a chicken Jenny describes as a "big personality" into Jenny's lap, something instantly snapped. Not only did Eloise become a pet but Jenny realized something. "I knew I'd never let her out of my sight again." Jenny and her husband started discussing house chicken options; she didn't want Eloise to live outside. Jenny already bolted to the backyard to see her flock as soon as she got home from work, and she and her husband occasionally hosted their chickens inside here and there. The move to permanent indoor residents didn't seem like much of a jump. "I started researching indoor chickens' needs, we set her up a space, and she loved it."

Bringing her chickens inside lit up Jenny's home. When she tells me the chickens filled her space with "laughter and antics," I don't doubt it, but I also suspect her brood is better behaved and more mellow than mine. When heavy downpours with high winds tear through southern Louisiana, I sometimes bring Joan and Amelia inside out of sheer anxiety, although I don't enjoy it. Joan's easygoing, but Amelia screams and pecks when she doesn't get the treats and attention she wants in the right combination (spoiler: there is no right combination). Jenny's the chicken equivalent of a mom with a "My kid's an honor roll student" bumper sticker, while my kid is the one starting fights over lunch money in the hallway. But for Jenny and her husband, it worked. They'd never been able to have pets before ("My mister is incredibly allergic to animals"), but chickens were different. They didn't cause a reaction. "We are so happy to be able to have our flock inside," she says. "It was really the missing link in our family."

Chicken people love their chickens. There is no way to convince a wide-eyed chicken person that however they're raising their chicken isn't the right way, whether it's in a coop worth thousands of dollars or in an indoor pen. "Now my entire life revolves around chickens, their health and happiness," says Jenny. She has three permanent house chickens and a brooder of six chicks. Her chimera chicken gets cannabidiol, or CBD, oil two times a day, and all of her chickens love to snuggle on the bed or couch with Jenny and her husband.

Jenny thinks chickens are an underestimated species and wants to normalize them as pets. "Perhaps it is because they are used for meat that they are dumbed down and considered filthy farm animals." Although they're dynamic and intelligent, it's supposedly easier to eat meat if it comes from something stupid and lowly.

"It is so sad to me the way chickens are treated in this world," Jenny tells me. "After having Ellie and Barth and Mia inside and seeing how loving and smart and playful and protective they are, I knew I had to be a voice for them." Jenny is a self-described house chicken advocate. She says her chickens love blankets, being inside, following her around, and watching movies. She makes them mini pancakes and features a #weirdthingschickenpeopledo post on her Instagram each week (@lipstick_and_chickens), which has over twenty-eight thousand followers. One week she tags me in a feature post for a video in which I'm vacuuming while Amelia perches on my shoulder, using her claws to cling to a dog chew toy that I've secured onto my body with a length of rope. Other features include a chicken wearing a taco costume, getting a bath, enjoying a beach picnic, and riding in a bicycle basket.

"The bond I have with them makes me want to fight to shed light on them as a species. They deserve love and affection and respect," Jenny tells me.

Through Instagram, Jenny highlights her chickens' personalities and showcases them in places that aren't coops. "I want chickens to be seen as equal to dogs," she tells me. "I get judged for having chickens inside. Why? People need to open their minds and see that it isn't an oddity. They might change their minds after visiting my home and meeting my flock."

On Christmas morning, walking to the coop in my parents' backyard with a bowl full of table scraps for the chickens, I think back to Jenny and how she never imagined she'd be a crazy chicken lady. I think about how Dad offered a loving yet resigned okay when I pitched the idea of driving cross-country with Joan and Amelia and about the emotions I could hear in his voice—exasperation, lack of surprise, recognition that I care about the chickens probably too much. As I follow the path along my parents' lawn to the L-shaped end of their backyard, eager to feed Joan, Amelia, Han Solo, Minty, Big Red, and Raquel with the salad, French bread crusts, egg shells, browned apple slices, and rice left over from last night's meal, I sing "Oh, What a Beautiful Morning" to announce my approach, smiling because the ladies hear me and cluster against the grate and rake their beaks across the wire mesh in anticipation, and I wonder, *Have I become a crazy chicken lady?*

6

Fowl Feast

WHAT CAN AND DOES GO WRONG

MY NEIGHBOR KEEPS chickens, although I mean that in the most base and lazy sense of the term, as in "I keep boxes in my attic." She doesn't maintain her flock with any degree of care or proficiency; rather, her chickens have a single, too-small coop that is ill fitted for predator protection and only remains marginally dry in the rain. She doesn't seem concerned with changing their water, cleaning their coop, or attending to the swampy mess of mud that their run eventually becomes. When I return home from California in January, I notice their conditions have worsened and numbers multiplied.

Although chickens can be a quiet form of resistance against the oppressive agricultural powers that be, they aren't, and shouldn't be, for everyone. A study by researchers at the University of California, Davis, published in *Poultry Science* in 2014, showed that most chicken owners know little or nothing about common communicable diseases that can infect and kill backyard hens, only 5 to

20 percent use a dewormer, and 41.8 percent of those surveyed didn't know that a vaccine for Marek's disease—one of the most common avian diseases, caused by a highly contagious strain of herpes—even existed.

From my side of the fence, I watch Taylor (not her real name) infrequently changing her chickens' water and occasionally throwing them grain or food scraps. Taylor is in her late twenties, with short hair and a stocky build. She has tattoos of things that represent her home state of Louisiana, like okra and the state's outline, on her arms and legs. Her chickens are rarely in her backyard. They run amok through the neighborhood, dodging cats and dogs and cars and kids in their daily fight for survival. At night, they roost in the fig and mulberry trees that overhang our fence. We find chickens wandering through our yard, scratching up our grass, laying eggs under a broken-down futon frame under our orange tree. They circle the apartment complex across the street and a few houses down like a roving band of land sharks, scavenging for scraps or security or both. Eventually the flock separates into two factions, each mastered by its own rooster, increasing the intensity and geographical range of their wanderlust. A third juvenile rooster bunks with the smaller sect, although I believe it's temporary and given enough time he will propagate his own clan.

There's a reason municipalities create laws curtailing chicken ownership. Most areas ban roosters, limit the number of chickens a single person can keep, create regulations about mandatory minimum coop size, or require permits. Lafayette doesn't do any of this. The laws here only state that chickens must be kept in their enclosure at all times and that the pen or coop cannot be within twenty-five feet of any building used for sleeping, living, working, or dining. Taylor doesn't seem to care. Her back fence is only about fifteen feet from

CHAPTER 6

her house. After almost backing over a chicken with my car, I start checking the driveway before I go anywhere. In the wake of her inability to keep the chickens contained, I worry about diseases.

After Kathleen of Moonwater Farm in Compton called the USDA to tell them her chickens were infected with Newcastle disease, the efficiency of their response was undeniable. "They swarmed the place," she told me. It was clear that their efficiency was born from experience and research. The USDA workers, with their airtight standard operating procedures, in from out of state, showed up wearing HAZMAT suits and sprayed themselves with industrial-strength disinfectant to avoid cross-contamination. They must have looked like aliens from another planet, donning protective gear to traverse the unfamiliar terrain that normally hosts farm camps and exists at the nexus of art and agriculture. "I called them myself to be a responsible owner. It's a virulent, almost plague-like disease," Kathleen said, and her actions were commendable. Not everyone would notify the powers that be.

Fortunately, Kathleen did. Newcastle disease harms the respiratory, nervous, and digestive systems of infected birds. It's so lethal that chickens can become infected and die before expressing any symptoms, which include gasping for breath, greenish diarrhea, tremors, and swelling around the neck and eyes. Newcastle disease doesn't pose serious health risks for humans, but it can jump to wild bird populations.

The team of professionals who came to decontaminate Moonwater Farm parked the truck close to the coop, put everything that could transmit the disease into plastic bags, and transferred the live chickens into trash cans hooked into carbon dioxide tanks. The birds were gassed.

Everything, including the workers' HAZMAT suits, went into black trash bags, which were loaded into the truck and taken to an incinerator. "They know what they're doing," Kathleen said. It was clear they'd done it before.

No one knows exactly where the Newcastle disease that infected Kathleen's chickens in 2002 came from. "One woman on the phone said they [the USDA] thought it came from Mexico, but they don't know that. We have a large Hispanic community, so it might just be the thing to say." The claim that Newcastle disease crossed the border is unsubstantiated.

America has had a longstanding connection between racism and animal agriculture. As early as 1909 in Chicago, biases against backyard chicken keeping were an indirect (or sometimes direct) way of attacking immigrant populations. That year, the *Chicago Daily Tribune* reported that "wherever immigrants who stand on the lower scale of industry live, thousands of chickens are being raised." There isn't evidence that minority populations contribute to poultry disease outbreaks, and more importantly, the animal agriculture system is based on environmental racism. Livestock operations are often located near or in communities that are predominantly home to people of color. This practice forces people with few resources to either fight back or to move away from dangerous public health conditions.

Most CAFOs (concentrated animal feeding operations) in North Carolina are in close proximity to predominantly African American, Latino, and Native American housing areas. CAFOs devalue property, pollute the air and water, and pose significant health risks. The racism surrounding urban chickens that Kathleen noticed lacks factual grounds, but the disadvantages that minority populations face from the pressures of commercial animal agriculture are well documented.

The Newcastle disease outbreak of 2002–3 that blighted Kathleen's flock at Moonwater Farm also infected commercial egg-laying facilities in San Bernardino, San Diego, and Riverside. By March of 2003, twenty-two commercial flocks had been contaminated, resulting in sizable financial losses for the poultry industry in that area. The disease spread through Southern California until September 16, 2003, when the California Department of Food and Agriculture announced that the quarantine could be lifted. By then the outbreak had cost more than $160 million and involved the euthanizing of over three million birds.

Fifteen years after the 2002–3 outbreak, Newcastle disease came back to Southern California. In April of 2018 the USDA confirmed a case of Newcastle disease in a small flock in Coconino County, Arizona, and by May there were 400 confirmed cases of Newcastle in Southern California, seemingly connected to the Arizona case. In the six months that followed, the outbreak was isolated to backyard flocks, but in December it jumped to its first commercial operation. Six large-scale operations in Riverside and neighboring San Bernardino were contaminated, and all 822,000 chickens in those facilities were euthanized. Accordingly, the 2019 Silicon Valley Tour de Coop was canceled in the name of biosecurity. The long-term effects of such outbreaks can be devastating for the poultry industry.

Vaccinating birds against Newcastle disease doesn't prevent it, but it does reduce the risks; unvaccinated flocks often experience a 100 percent mortality rate. There is no known cure. In a 2014 survey by researchers at the University of California, Davis, of backyard chicken owners, only 30 percent of respondents knew about Newcastle disease. This limited knowledge led the researchers to conclude that survey participants expressed a "deficiency in knowledge" that could potentially cause their mismanaged backyard flocks to contaminate commercial flocks. Just over 41 percent of the survey participants made $100,000 or more annually, suggesting that the problem isn't about access to information or resources.

I can't tell if Taylor lacks information, resources, or both. As the chickens flee her yard and seek sanctuary from the street, they end up in our backyard, which becomes laden with foreign feces. The dogs roll in it, the humans step in it, and we warn houseguests against walking barefoot on the lawn. We have fresh grass, clean water, and a constant supply of chicken feed, so I can't blame her chickens, but I can blame Taylor. *Salmonella*, I think, glaring over the fence at her as she weeds her garden beds, plotting the indictments I would tattle to an Animal Control officer given the chance. *They're feral*, I'd say. *They roost in the mulberry tree at night and come in our yard*, I'd tell them. *Help me*, I'd beg. *The dogs are going to get sick. The humans are going to get sick. Her chickens are sick—I'm sure of it.* One afternoon I watch one, her left foot double the size of her right, swollen with some terrible infection, wandering around in the gravel lot across the street. I try to catch her—and somehow fail despite my clear advantage.

Salmonella is one of the top concerns related to backyard chickens, at least according to the powers that be. In 2018 the Centers for Disease Control released a report that accused backyard chickens of causing a nationwide salmonella outbreak, which resulted in 334 cases of infection and 56 hospitalizations across forty-seven states (fortunately, this did not cause any deaths). During interviews, 63 percent of those infected reported handling chicks or ducks the week before they got sick. Keeping backyard chickens involves an inherent risk of disease; you're dealing with livestock in an urban space.

CHAPTER 6

According to Chicken Mike and Nicole Graham, the risk of contracting an infection from backyard flocks is blown out of proportion. When I visited them in Houston, we went to one of their clients' houses—a multimillion-dollar brickwork mansion with sweeping lawns and an expansive driveway—to clean the custom chicken coop in the side yard. It was a two-story enclosed coop and run, large enough to stand up in, with a raised nesting area. The back door opened to the nest boxes for easy egg access. After removing all the old bedding, hosing out the concrete-bottomed run, and disinfecting everything with a diluted vinegar solution, Nicole—wearing a tank top, jeans, and cowboy boots—crawled into the top level of the coop to spread clean pine shavings. The client's boyfriend came outside in a suit and tie.

"Did you see that news article?" he asked, referring to the CDC's report. Chicken Mike, already stripped down to his white undershirt from physical labor in the heat, shook his head.

"Are they safe?" the boyfriend asked, referring to the chickens' eggs. Mike assured him they were. The suit handed me his phone with the article pulled up. "Read this," he instructed. I fantasized about and dreaded the possibility of dropping his phone on the expensive-looking brickwork driveway.

Later over lunch, Nicole and Chicken Mike told me that backyard chickens often get a bad rap for spreading diseases, but the number of illnesses they actually cause pales in comparison to those catalyzed by industrial poultry operations. "Chickens in factory farms are unhealthy," Nicole said. "Of course they're always getting sick. Of course they get people sick." Nicole argued that the chickens raised in factory farms who "live tortured lives" are overfed antibiotics and bred to grow faster than their organs and skeletal frames can support. "Where in nature do animals grow that fast? How is that natural? How is that healthy?"

The idea that news about salmonella from backyard chickens is political propaganda aimed at protecting the poultry industry, not health PSAs from objective onlookers, might sound radical, but others feel the same way. Jenny, the self-proclaimed house chicken advocate, also thinks the risk of salmonella is blown out of proportion. "People want you to eat meat from factory farms, so articles about salmonella are pushed out so you'll get scared of getting sick." The point is to drain the backyard chicken movement of momentum.

"There could be one sick chicken, and they'll send reporters," Chicken Mike said after the suit went back inside. He argued that instances of illness around backyard flocks are overreported while those that happen in commercial facilities are predominantly ignored.

It's understandable that the commercial egg industry would have a vested interest in raising the alarm on backyard chickens. If people care for the chickens in their yard, they might become more concerned about the animal welfare abuses of large-scale agriculture or realize that the bright orange yolks from their chickens' eggs that hold their shape in a frying pan are what yolks are actually supposed to look like, not the flaccid, pale yellow mess found in grocery store eggs.

If any chickens are going to be the harbingers of disease, it's Taylor's flock. I plot their demise, eyeing them through the gate as they dig holes in the brush alongside the road. I envision a rooster battle to the death that never comes, placing secret bets on the quieter, smaller rooster. I hope desperately that he somehow tasks himself with executing the louder, larger rooster, who begins crowing with gusto around midnight and choruses until 8:00 a.m.

CHAPTER 6

"You know," I tell Taylor through the fence one day, "if you'd like, I can show you how to kill him." I do not mention that I would view this act as two parts self-interest and one part mercy.

"I might be more open to that later," she says, rubbing her hands, dirty from gardening, on her Spandex shorts. "I think right now I have some other projects going on. And my dad knows how to butcher chickens—I can ask him."

"Okay," I reply. Normally I'm more assertive than this. I've asked Katie more than once if she thinks I'll be providing an honorable neighborhood service or just starting an all-out war with Taylor if I slaughter the rooster in secret. If I wasn't so terrified of Taylor, I suspect the rooster would be gone by now. She manhandles her chickens, yells at her dogs, and has had screaming fights in the street. I try to balance the probability of a communicable disease originating in her backyard with my fear of confrontation. I consider calling Animal Control, palming this last-ditch effort like a balm on a wound that just won't heal. A second coop materializes and in my cowardly desire to avoid the issue, I placate myself into believing this could be a turning point. Although she works in a local feed store, she seems to either not know or not care about her chickens' basic needs.

Taylor talks about and treats her chickens like extensions of her garden, productive units to plant and all but leave alone. Once, when she told me how many eggs she got per week, a number that seemed far too low considering her large flock, I suggested that maybe something was interrupting their production—stress, not enough food, the lack of nest boxes, or even disease. "I get enough eggs," she said, and the conversation was over.

As long as they meet her needs, Taylor doesn't appear to consider theirs. When Taylor isn't home, I throw her chickens grain over the fence.

Because the CDC blames the recent uptick in salmonella outbreaks on urban chickens, it's easy to see why some people are worried. Headlines can make the current state of affairs sound like a full-scale salmonella apocalypse: "Multistate Outbreaks of *Salmonella* Infections Linked to Contact with Live Poultry in Backyard Flocks," "Why Backyard Chickens Are a Health Risk," "Over 200 Salmonella Infections Linked to Backyard Chickens, CDC Warns," "Backyard Chickens Carry a Hidden Risk: Salmonella," "CDC: Backyard Chickens Are Giving Their Well-Meaning Owners Salmonella." The message is clear: backyard chickens are filthy, and we're all going to die of food poisoning if we keep them around.

The truth is a little more nuanced. In "Don't Play Chicken with Your Health," the CDC reported that between 2000 and 2018 backyard chickens were associated with 76 salmonella outbreaks in total, resulting in 5,128 illnesses, 950 hospitalizations, and 7 deaths. Although this is devastating, commercial poultry is also suspect. In the United States there are about a million cases of salmonellosis annually, resulting in approximately 400 deaths and around 20,000 hospitalizations: about a third of those illnesses, deaths, and hospitalizations are attributed to commercial chicken, beef, pork, or lamb. Chicken is responsible for causing more foodborne illnesses than anything else Americans eat, and in 2018 alone there were 866 cases of salmonella from commercial poultry meat and eggs.

Back in 2015, the USDA's Food Safety and Inspection Service began a program of testing raw chicken and established a "maximum acceptable" rate for how much raw chicken contaminated with salmonella a slaughterhouse could churn out: 9.8 percent of broiler carcasses, 15.4 percent of chicken parts, and 25 percent of comminuted (minced) chicken could be infected with salmonella. Facilities are placed in categories: 1, 2, or 3. Slaughterhouses that perform at or

CHAPTER 6

better than 50 percent of maximum acceptable standard are placed in category 1, those that meet the standard are placed in category 2, and those that fail it are placed in category 3. Between May 2018 and May 2019, an average of 32.08 percent of the 187 tested facilities fell into category 2 or 3.

Multi-drug-resistant salmonella outbreaks linked to commercial poultry are not uncommon, and outbreaks can be deadly. When people get sick from factory-farmed chickens, they've bought a product they assume to be safe. When illnesses occur from backyard flocks, it's often from young children kissing, cuddling, and snuggling their pet chickens or coming in contact with feces and not washing their hands.

The chicken industry also has an important tool critical to human health and absent from most backyard flocks: antibiotics. The potential for antibiotic-resistant bacteria outbreaks is arguably one of the largest threats to public health posed by industrial agriculture. In 2016 Dr. Margaret Chan, director-general of the World Health Organization, stated that antibiotic resistance is a global crisis that constitutes "a fundamental threat to human health, development, and security." Approximately 80 percent of all antibiotics consumed in the United States do not go to people; they go to the animals people eat, and it is not uncommon for animals in factory farms to be maintained on a low dose of antibiotics for their entire life. Antibiotics are used to promote growth in chickens and to keep them relatively disease-free in cramped quarters, but antibiotics were not always central to animal agriculture.

During the outbreak of World War II, the products originally used to feed single-stomach animals like chickens and hogs were in short supply. It was common knowledge that chickens grew faster and healthier when they were fed diets rich in animal proteins. These were usually delivered in the form of fish meal, cod liver oil, or even rendering plant by-products, but after the United States entered World War II these products were scarce. The search for an alternative protein source began.

In the late 1940s, a solution seemed to be on the horizon. The pharmaceutical company Merck discovered how to synthesize low-cost, highly concentrated vitamin B12, an important compound found in animal proteins, from vats of fermented microbes. In 1950, curious as to why chickens who scavenged and dug through manure laid more eggs and suffered from fewer illnesses, scientists at Lederle Laboratory conducted a series of studies and discovered that the bacteria and microbes in manure produced quantities of a B12-like substance. In an experiment, chickens were fed a version of B12 created with the antibiotic Aureomycin.

The chickens fed the new B12, laced with small doses of antibiotics, grew faster than those who didn't receive the concoction. Farmers across the nation began giving chickens B12 with antibiotics instead of the animal proteins of the past. The time it took for birds to reach slaughter weight dropped, as did the price of chicken in the checkout line. Today, animals are given nontherapeutic doses of antibiotics in their feed and water at the rate of a tenth to a hundredth of what would constitute a medical dose, or one provided to cure a bacterial infection.

Antibiotics have an important place in society. Under correct circumstances and responsible usage, antibiotics save lives. However, feeding antibiotics to livestock animals to make them grow faster and prevent them from being overtaken by the diseases that would otherwise result from the crowded, unsanitary conditions they experience poses a huge risk to human health. In *Big Chicken: The Incredible Story of How Antibiotics Created Modern Agriculture and Changed the Way the World Eats* journalist Maryn McKenna wrote that "nearly two-thirds of the antibiotics that are used for those purposes [faster growth and disease prevention] are compounds that are also used against human illnesses—which means that when resistance against the farm use of those drugs arises, it undermines the drugs' usefulness in human medicine as well." Antibiotic resistance is a natural process whereby bacteria become invincible to the compounds designed to annihilate them. A specific bacteria can become resistant to antibiotics in a variety of ways, including changing its membrane structures so the drug cannot penetrate the cell walls, changing the structure of the surface so the drug cannot bind to it, altering the active molecules in the antibiotics, or even pumping the drug molecules out of the cell. When bacteria successfully mutate in a way that allows them to fend off antibacterial agents, they have an advantage over non-resistant bacteria. The resistant ones grow and divide while the non-resistant ones are killed.

It is possible to mitigate the risk of antibiotic resistance by lessening the exposure bacteria have to antibiotics used to counter them. "What slows the emergence of resistance is using an antibiotic conservatively: at the right dose, for the right length of time, for an organism that will be vulnerable to the drug, and not for any reason," writes McKenna. "Most antibiotic use in agriculture violates those rules. Resistant bacteria are the result."

Currently, a number of antibiotics are used in livestock feed. Animals are given chemically synthesized antimicrobial agents, antiprotozoal agents, and microbial-based antibiotics, including chlortetracycline, procaine penicillin, tylosin, oxytetracycline, bacitracin, and oleandomycin, to name a few. Although feeding animals continual low doses of antibiotics is common, increasing the dosage during times of disease or stress is also an accepted industry practice. According to a study published in *Clinical Microbiology Reviews* that focused on the impact of antimicrobials given to food animals on human health, "Among the various uses for antibiotics, low-dose, prolonged courses of antibiotics among food animals create ideal selective pressures for the propagation of resistant strains."

Although there have been bans on giving livestock nontherapeutic doses of antibiotics in Europe, the United States does not have any such restrictions and continues to debate the connection between antibiotic resistance and antibiotics in animal feed. By some estimates, antibiotic resistance will cost the world $100 trillion and cause ten million deaths annually by 2050. The United States has failed to adopt the "precautionary principle," an idea that originated in the 1970s and directs policies concerning public health in the European Union. The precautionary principle "asserts that parties should take measures to protect public health and the environment, even in the absence of clear, scientific evidence of harm. It provides for two conditions. First, in the face of scientific uncertainties, parties should refrain from actions that might harm the environment, and, second, that the burden of proof for assuring the safety of an action falls on those who propose it." The cost of meat could rise if producers in the United States avoid using antibiotics in animal feed. In 1980 the National Research Council Committee to Study the Human Health Effects of Subtherapeutic Antibiotic Use in Animal Feeds suggested that poultry houses would have to create upgraded maintenance and environmental control systems as well as increased automation to limit human contact, although today antibiotics are still used in poultry feed and these systems are not utilized to avoid those additives. It is unlikely that the price of eggs and chicken would remain the same if antibiotics were taken out of the picture.

CHAPTER 6

Antibiotic resistance can be transferred to slaughterhouse or poultry farm workers, veterinarians, and others who come in contact with chickens or their feces. While this might not seem like a significant number of individuals in our population, workers and their families can act as a point for resistant bacteria to transfer into society through hospitals and individual contact. It is also possible that resistant bacteria can come home with consumers from the supermarket. The risk of consuming resistant bacteria on meat is well documented, as are cases of antibiotic resistant infections in human populations beginning in factory farms.

It's a grim possibility that awaits our society if some of the predictions concerning the rise of antibiotic resistance are true. The idea of ten million deaths annually from resistant strains of bacteria is staggering, especially considering the ways animal agriculture has contributed to, shaped, and sharpened the risks that face us in the future.

Although I want to condemn my neighbor for the way her backyard chickens could spread diseases as they wander wild through the neighborhood, her individual capacity for creating a dangerous environment is limited. The poultry industry's capacity for this, however, is staggering. Taylor's chickens will not cause a widespread, multi-drug-resistant outbreak, although the industry is creating the conditions that might contribute to something detrimental.

Although chickens are clearly not for everyone, I feel far more comfortable with the idea of practicing the typical measures of biosecurity—like washing my hands and not kissing my chickens—to mitigate health risks than I feel with the global threat of antibiotic resistance. Yes, there are bacterial outbreaks from backyard flocks, but chickens in backyards are not commonly fed diets that continually expose them to low levels of antibiotics, allowing the bacteria inside their bodies to build up immunities to the antibiotics and develop ways to render the drugs inert. The chickens who live in backyards are not as crowded and filthy as the chickens in factory farms, and although these small measures are hard to visualize, the effects are important.

I think of organically raised factory-farmed chickens and how they do not receive antibiotics even if they need them, or how, if they do, they can no longer be marketed as *organic*. I remember Isabelle, whom I visited during Silicon Valley's Tour de Coop, and the one-eyed chicken who was rescued from an organic egg farm. Curry didn't get medication for an infection, and Isabelle told me missing eyes were relatively common in the chickens she helped rescue from the farm; a third, she said, were missing eyes.

There are degrees of risk and failure, it seems. It is clear the precautionary principle developed by the European Union would benefit American consumers and, arguably, the world, since antibiotic resistance isn't likely going to be a problem in the United States alone. This is a global threat. Organic chickens who aren't given antibiotics might not get the care or compassion they need. Backyard chickens, if their owners are irresponsible, might pose contained health risks and might not get the care they require. My goal is to fail less—to have healthy chickens, provide them with the care they deserve, practice basic sanitary measures, not consume meat or eggs that come from birds who have been continually fed antibiotics, and not pay money for goods that come from an industry that does.

A couple of days after the rooster conversation with Taylor, I find a rat in my chicken run. When I try to coax him out with a broom, he jumps, jerks, and wobbles as if he's been poisoned. I eventually manage to push him out of the yard.

CHAPTER 6

When I pass the fence to Taylor's, I notice a rotting rat carcass by one of her henhouses. Maggots crawl in and out of eaten-in dimples in its face flesh. It smells like it's been there for a few days.

Still holding the broom, I knock on her front door. "I found a rat that looked poisoned." She extricates herself from the house, carefully opening the door large enough so she can fit through. Her two dogs bark from the other side. "Did you put out any poison?" I ask. The dogs jump against the windows, rattling them.

"No, I didn't. I don't think the neighbors to either side would, but I can ask. Thanks for letting me know."

"No worries. I just wanted to tell you that there might be poison around."

"I'll keep an eye on the dogs," she says.

"Cool," I say. "Also . . . um . . . when I was pushing the rat out of my yard, I noticed a dead rat in your chicken run. It's by the coop."

"Oh, really?"

"Yeah, maybe check it out." I wanted to tell Taylor that I think her chickens should be taken away from her and that she's creating a public menace. I'm almost positive her limping, swollen-footed chicken has bumblefoot, an infection caused by bacteria entering a small wound, and I'm terrified my chickens will catch it.

When I don't leave, she asks, "How's the rooster?"

"Well, you know, he doesn't really bug me," I lie, "but I think he keeps Katie up."

Taylor says she'll take care of him, and a week later he's gone. Instead of killing him, she gave him to a woman who frequents the feed store where she works. I hope he is happier, although I'm sure his new owners aren't.

Diseases aside, it can be hard for chicken owners to keep their hens for the long haul. Chickens start laying eggs when they're between four and six months old, and although their production begins to wane after eighteen months, their laying abilities continue to decrease and are significantly affected when they are around two to three years old. The problem? Chickens can live for eight to ten years, and sometimes double that, so owners need to commit to housing, feeding, and cleaning up after a freeloading bird who isn't providing breakfast, slaughter her (an unpopular option for those who've named their chickens, posted pictures of them in hats on Instagram, and house them in expensive custom coops), or rehome her.

While the Humane Society of the United States supports reducing animal suffering and recognizes that when families keep backyard chickens they likely aren't buying eggs from hens raised in factory farms, they also recognize that rescues can be overburdened with unwanted birds and that roosters are sometimes either killed or abandoned. The pastoral whimsy of baby chicks can be intoxicating, and caring for a hen in exchange for fresh eggs makes sense, but when chickens all but stop laying eggs, most people's interest wanes. When I started finding chickens to eat on Craigslist, they were almost exclusively roosters and old hens who had stopped laying. No one wanted to kill their pets, but neither did the environmentally responsible locavore owners want to keep chickens who weren't laying eggs.

If someone can't find a good home or soup pot for their chicken, the hen in question can end up abandoned on the streets or surrendered to a shelter. In the past twenty years, backyard chickens have been showing up in shelters at unprecedented rates. Since Chicken Run Rescue in Minneapolis opened in 2001, they've taken in 1,158 surrendered, stray, or abandoned chickens. According to a 2013 article published in *Forbes,* the year they opened, owners Bert and Mary Clouse received six calls from people trying to rehome their urban

CHAPTER 6

farm animals; by 2012, that number skyrocketed to five hundred. During an interview with the *Minneapolis Star Tribune*, Mary described the situation: "It's like watching a train wreck in slow motion." Some of the chickens they've housed have special needs, like the neglected chicken that lost its feet to frostbite—the result of inappropriate housing combined with a Minneapolis winter—or the chickens with reproductive cancers caused by constant egg laying.

Although offering to execute Taylor's rooster before it was rehomed might seem extreme, there's an almost unavoidable level of violence built into the backyard chicken system, regardless of how old birds are treated. The idea of voting with your dollar when you go to the grocery store—do you support the cheapest, most convenient option or an environmentally sustainable, responsible one?—can easily extend to backyard chickens, but not in the way you might think. Slaughtering an unwanted rooster means dealing with a chick that wasn't sexed properly or came from a source that doesn't perform the process.

Hatcheries provide sexed chicks, which have been determined male or female at birth, so buyers can choose egg-laying hens or roosters. Most backyard chicken owners don't want the boys, and some cities don't allow them; demand for the ladies always is, and probably always will be, higher. Male chicks are often considered an unfortunate by-product. When I visited Isabelle at Clorofil, we talked about this inherent violence toward and devaluing of male chicks. "When people buy baby chicks from a hatchery, they don't realize that the little brother has been killed," she told me. "They say, 'Well, it's okay, my hen. It's a happy egg because it's in my backyard.' Well, there is death from the beginning."

We didn't discuss the details, but the two most common methods for culling unwanted roosters are suffocation from carbon dioxide or maceration, a process that involves a machine akin to a woodchipper. Day-old male chicks are ground up while they are still alive, and this happens to heritage chickens and factory-farmed chickens. For every egg-laying hen, there's a male chick that was killed.

Even when people buy chickens to raise in their backyard, they are paying hatcheries that are tied to the commercial poultry industry. I know, I did it. "I think even backyard chicken people don't realize that," said Isabelle, and she's right. If people want designer or heritage chickens, it's almost impossible to avoid hatcheries unless they're incubating their own fertile eggs. Buying chickens at a feed store doesn't bypass the hatchery system either, because that's most often where they're delivered from. I could've adopted chickens someone didn't want, but I was hoping to raise them for eggs or meat first and for pets second.

Isabelle compares buying chickens from a hatchery to buying a puppy from a breeder. "And people don't think back even further to the lives of the parents. Who is laying those eggs that will hatch?" she asks. "Where are those parents living? Well, they live in conditions I cannot describe to you, in a factory farm. It's a breeder farm, and they are just bred and selected really young, like every year, culled based on their features, so they can pass the right genes to the offspring, and that part is hidden so well that very few people bring that up right now."

CHAPTER 6

As a teenager, I was informed that if a cop showed up to the door I was under no obligation to let them into the house without a warrant, that I really didn't need to roll down my car window farther than it would take to slide an ID out, and if I was questioned by an officer on foot it was best to ask, "Am I being detained?" to establish whether or not I was free to go. I had friends who read political science and participated in direct action. In college, I knew people who would make me take the battery out of my phone before coming into their dorm room and handed out pamphlets with headlines like "Know Your Rights." For these reasons, opening the swinging gate to our backyard and welcoming in an Animal Control officer makes me nervous.

I didn't call Animal Control, but apparently in the past few days, "several" of the neighbors had. The officer parked her car by our back driveway, got out, and taped a yellow court notice to Taylor's door. From inside, Taylor's dogs jumped against it, their nails clattering against the wood. Katie and I had been discussing the possibility of calling Animal Control, but we hadn't. There was a sense of relief in the fact that someone else had. Katie watched the scene unfold from the back patio. The officer walked around Taylor's front yard—as much as she could from the street—and peered along the fence line to the backyard.

"Did you see the chicken in the street?" the woman asked Katie.

"They're always in the street," she responded.

After Katie ducked inside to call for me, I came running, elated that perhaps this was the end of Taylor's antics. I just didn't expect I'd be letting the officer into our yard.

She is a parched-looking lady in her fifties with a cigarette smoker's voice, raptorial hands, and a lean, crow-like body. Her job? Collect evidence. She walks down the street, leans over Taylor's gate again, and takes pictures of the wandering chickens and squalor in the yard. The last straw, if I had to guess, was the recent accidental release of a bunny Taylor adopted to breed for meat and fur despite never having raised rabbits before. The rabbit was only in the yard for a day before it made its great escape, and in the afternoons, Taylor could be seen running through the street with a giant fishing net, the black-and-white rabbit growing more feral and evasive by the day. That is, until she gave up and let it run wild. My friends and I tried bribing it, then chasing it, but we were about as proficient in our efforts as Taylor.

When the officer can't get a good look at the chicken from the neighbor's yard adjacent to Taylor's from the street, I let her in. After squatting down, failing to see the chicken that had been there moments before, and walking farther into my backyard for a better view, she spots my coop.

The fact that I volunteer to let her come into our yard speaks to the bind I feel I'm in, but I am delirious from months of rooster-interrupted sleep, waves of rage, and the hours I'd lost chasing the rabbit and the limping chicken before him through the streets. I will do anything.

"Is that," the woman asks, taking a sharp breath while she points a predator finger at the blue-painted structure, "a chicken coop?" I glance at Katie, who gives me a look of horror. I wonder if I just ruined my backyard chicken project by inviting this lady into the yard, and I can tell Katie's thinking something pretty similar.

"Yes," I hurry to answer, "that one is ours, but it's twenty-five feet from all houses, they are not allowed to run free, they remain in their enclosure all day, and we keep them locked in the coop at night." I regurgitate everything I memorized from the municipal codes. "Plus," I add for good measure, "we only have two."

CHAPTER 6

"This isn't legal," the woman says point-blank. Her words slap me. I run through all the possibilities—bunk Joan and Amelia with some friends until this blows over, actually get rid of the chickens for real, or move them inside, cover the coop with a blanket, and say it's just some kitschy kind of yard art.

Katie, standing beside me on the lawn, looks stricken. I feel like an idiot. Is uninterrupted sleep, not worrying about disease transfer to my birds, and a clean lawn worth giving up my chickens?

"Excuse me?" I ask, shocked.

"It has to be twenty-five feet from anywhere used for living, eating, or playing. See that yard?" The chicken run shares a fence with a two different neighbors' yards. "That's not twenty-five feet."

"No," I counter more aggressively than necessary, "the law specifically says a *building*. If they can't be within twenty-five feet of a space used for play, how could you even have them in your own private backyard? Isn't this a space used for play?"

"That's why you move to the country," she says tersely. Another option flashed through my head with chilling force and precision: knock the woman out, throw her in the backseat of her car, and drive it into the Atchafalaya Basin.

"That doesn't even make sense."

After a brief scrimmage about what is and isn't legal, I pull up the parish's municipal codes on my cell phone. "Look," I tell her, not bothering to mask my self-righteousness. "There." I point to the screen.

"Well, all right," she says. "They changed recently and I better check them again. But I'm not coming after *you*."

Damn right. I normally believe that The Man should keep to himself and let people do their own thing. Most days, I find the chaos of urban chickens utterly charming, but right now I want a bureaucratic hammer to come down on Taylor, complete with regulations, prohibitions, a trial, and maybe even penalties in the way of exorbitant fines and/or jail time.

Clearly, I'm not as counterculture as I'd like to think. I hailed an Animal Control agent, ratted out my neighbor, and did it because I am tired of finding chickens who aren't mine on the lawn and checking under the car before backing up. I just want Taylor to pay for her trespasses, which is not to say I don't have any of my own.

One night, after a whole lot of complaining with Katie, some of Taylor's chickens fly into the mulberry tree to roost. We watch them as they flap in the branches before settling down. The day before, when I passive-aggressively suggested to Taylor that perhaps I could help her keep *her* chickens out of *our* yard by clipping their wings, Taylor assured me she had already taken the time to do so. "It was hard to catch them," she lied (even then, I knew with the rage of unfounded conviction that it absolutely had to be a lie), "but I got them all."

Katie and I are both fed up with the chickens roosting in the trees and with Taylor in general. Emboldened by IPA, under the cover of darkness, I stomp through the grass, upturn a bucket to create a makeshift stepstool, and pull a chicken from the branches. The others flee to Taylor's yard. *At least that's where you belong*, I seethe. Frustrated and tipsy, I return to the back patio, sit next to Katie, and check the chicken's flight feathers. They are completely intact.

"Oh, hell no," I spit before storming inside, heading to the kitchen, and grabbing a pair of scissors, all with the bird tucked under one arm and Tashi and Atlas marching behind me in a procession of wagging tails. Normally I would ask Katie if this is a good idea. Normally I'd trust her judgment if I felt like I was being unfair and perhaps a bit extreme, but I have the scissors and I know what I think needs to be done. When I get back outside, I smack down on the lawn chair, elongate the chicken's feathers with one hand, and clip her wings with the other.

"Do you think she'll find the feathers?" I ask Katie in a passing moment of self-doubt. Clipped feathers scatter our covered porch.

"No way," she says, shaking her head. I make a mental note to clean them up and throw them in the compost anyway, just in case.

I finish clipping the wings and make sure it is complete. I leave a few of the flight feathers near the wingtip so my hack job is hidden when the chicken's wings are in their normal resting position. After getting back up on the bucket, I drop her over the fence. She lands with a soft plop.

The next day she is one of only two birds still in the yard; the other is the young rooster, trying to get on top of her.

About a month later, when Katie and I are outside having coffee one morning, we notice three baby chicks following the chicken with clipped wings. The four of them wander into the street, loop around our driveway, and then head through the neighbor's yard. The mother chicken kicks up leaves as she pecks for bugs.

Neither one of us laughs. As Katie and I witness this miracle of creation, her with a perspective that contains far less culpability than mine, I swear out loud.

"I just made it worse."

7

Eating Bugs for the Environment

THE CHICKENS AND I SHARE A MEAL

I'M IN THE local pet store, shuffling through the insect fridge near the reptile supplies. The goal is to locate the two most recently dated cups of live wax worms. Some of the containers don't have dates printed on the sides, so I line them up on a nearby shelf and peel back the white lids. The wax worms, which aren't really worms but rather the larval stage of wax moths, writhe like irritated yet sedated sleepers when I poke them. They are about three-quarters of an inch in length, packaged in pine shavings and dulled by the refrigerator's cold. I sift through the cups with my fingers, trying to count how many blackened, shriveled carcasses are in each. There are reputable, human-food-grade suppliers of insects, but the pet store seems like an easy option. I convince myself that since the bugs are, hopefully, going to breed it doesn't necessarily matter what my future meal's parents were eating. Plus, a lot of companies sell already-dead insects, which isn't what I'm looking for. Even the live wax worms, the healthy and plump ones, don't look particularly appetizing, although I know better.

CHAPTER 7

Eating insects is surprisingly good for you, but this isn't the only reason to do it. Take mealworms for instance: the amount of unsaturated omega-3, protein, and vitamins in mealworms compares with that of wild-caught fish, and they have more omega-3s than beef or pork and more vitamins (besides B12) but slightly less fat than beef. Of the six essential amino acids, mealworms have high concentrations of four. Consider crickets, which have the same amount of protein as pork. Although there is some debate about how well humans can metabolize insect protein, a conversation rooted in ambiguity about whether or not mammals produce an enzyme necessary for breaking down the exoskeletons of insects, most of this research has been done on people from Western nations rather than individuals from cultures that do eat bugs. Most primates, however, still retain this enzyme. Cooking and chewing insects help break down the exoskeleton, and according to the United Nations, nineteen hundred species of insects are edible and highly nutritious sources of protein, healthy fats, fiber, vitamins, and minerals.

For some of the same reasons they're good for us, bugs are good for chickens. The amino acids, vitamins, and minerals found in insects help promote healthy yolks and hard shells. Most commercial chicken feeds rely on soybeans or fish meal for protein, but new research found that replacing commercial broilers' conventional feed with an alternative heavy in black soldier fly larvae didn't affect feed intake, average daily body weight gain, the smell or taste of the meat, or the feed conversion ratio. The replacement did, however, decrease feed costs by almost 20 percent and represents a much more sustainable alternative to soybeans or fish meal. Adding insects to chickens' diets during certain times, like when they're molting or if their shells are thin, can give them the nutritional boost they need.

Although the grubs look like blind maggots wearing little leather flight caps since their hard-looking black or brown heads are dark like the tips of their small, searching arms, they are delicious. Yes, eating bugs seems gross, but once you try them, they aren't that bad. The little caterpillar-like larvae crawl by inching their squishy, slightly articulated bodies. They are high in fat, taste a little bit nutty, and are just slightly sweet, which makes sense: in the wild, wax moths lay eggs inside beehives, leaving their protégés to gorge themselves off of honey and beeswax as they develop, much to the chagrin of beekeepers (it's no surprise they're also called "bee moths"). From my personal experience, I find they should almost always be sautéed with olive oil, onions, peaches, and garlic and wrapped into warm tortillas.

But before they can become tacos or chicken treats, these wax worms need to breed. I want to grow my own wax worms, hence the pet store container shuffle, because it doesn't seem practical to continuously buy them. I also want to raise enough for my chickens to enjoy an extra kick of protein to supplement their dumpster diet. Joan and Amelia are shamelessly demanding when it comes to dried mealworms, and although companies like Grubbly Farms produce high-quality treats for chickens, I'd like to opt out of the capitalist system on this one. It seems only fitting that if I am to consume and feed my chickens grubs, I understand some of the mysteries of their life cycle, control what they eat, and avoid buying treats that come in plastic packaging.

I take the wax worms home, carefully handpick them from the pine shavings, and drop them into a terrarium I prepared the day before. I mixed baby cereal with honey and corn syrup to make their two-in-one food and bedding combination, lathered it on the bottom

CHAPTER 7

of a five-gallon tank, then balled up a few wads of parchment paper. The worms, in theory, will eat the sugary mash, grow, crawl onto the parchment paper and spin themselves into cocoons, then hatch as moths, lay eggs, and die. The eggs will then hatch and—voila!—tacos and chicken treats.

The worms appear content. I put them on top of the refrigerator (they prefer warmer temperatures, but the internet assures me that even if they're kept at room temperature they will breed; it'll just slow down the operation), place a cloth over the tank to keep it dark, and leave them there for a couple of days. At first they appear fine, and when I peek under the cloth they're squirming along the cereal, and I can only guess that they're eating, although I'm not sure what their motions actually indicate. The next day they look a little slower and are still moving, but less so. On the third morning I show up in Katie's doorway, clutching the tank to my chest.

"Katie," I tell her, "the worms are dead."

"What do you mean, dead?" she asks, sitting up in bed. Beside her, Tashi snores and sleeps, oblivious to my dismay.

"I mean, they're just not moving." I am astounded that after a mere seventy-two hours I managed to kill larvae that were literally existing in a half-frozen state suspended in pine shavings and the carcasses of their fallen kin, when all I did was heat them up and give them food. In a sad attempt to delay the inevitability of scraping dead caterpillars and the spackle-like paste of cereal, honey, and syrup, I leave the terrarium in my closet and go camping for the night.

ᚲ ᚲ ᚲ ᚲ ᚲ ᚲ ᚲ

Other people seem totally capable of raising bugs. Angela, who raises mealworms for her chickens, created her own mealworm farm at home. She's a do-it-yourself kind of person who believes in real food and mixes her own chicken feed. "It really just made sense for me to start a mealworm farm," she tells me. To build a system and fill it with a thousand mealworms cost Angela a total of thirty-five dollars, plus a few bucks for food and bedding. "Each beetle can lay three hundred to five hundred eggs, which translates into a lot more mealworms than what I buy for forty dollars a bag." She feeds her mealworms organic, high-quality foods so she knows that her flock is getting the best. "I spoil my flock more than my kids," she admits.

Not only is it cheaper for Angela to raise her own mealworms, but it's potentially safer. A lot of mealworms sold in the United States are grown in China. Mealworms are surprisingly resilient and can handle consuming a whole host of strange products. They can even stomach eating exclusively Styrofoam for a month. The chemicals used to make Styrofoam, like formaldehyde and styrene (which have both been linked to cancer), can leach out of their bodies. There are safety concerns among some backyard chicken keepers about mealworms imported from China—most notably, that the dried grubs contain harmful chemicals and questionable substances. There's not a lot of testing for this, but these concerns do have valid components.

In some ways, consumers don't actually know what goes into prepackaged dried insects. If someone wants more control—if they want to have the level of autonomy to make those decisions and remove the risk of exposing their chickens to chemicals—the options are to research the heck out of the companies selling mealworms or grow the bugs themselves, like Angela.

While she says it's not difficult or intensive, according to Angela the most time-consuming aspect of raising mealworms is moving them into the correct drawer so the beetles don't eat the pupae. "But really, moving beetles and pupae is really just a few minutes a day." The labor-to-cost ratio is pretty attractive, but I'm daunted.

Phase two is crickets. If I can't keep caterpillars alive long enough for them to turn into moths, perhaps I can handle something that doesn't have to undergo metamorphosis. Mealworms are out. I order five hundred crickets online and hope for the best.

In theory, crickets won't be too much of a challenge. Austin Miller, founder and chief cricket guy at Craft Crickets, assures me that "insects are nutritional and easy to breed." Craft Crickets is Oregon's first licensed cricket farm (yes, that is a thing) and is housed behind a Burger King in Eugene, proving that crickets can be grown practically anywhere. While most cricket breeders raise bugs to fuel pet lizards, the insects at Craft Crickets are exclusively for human consumption. They come flavored with cinnamon sugar and paprika. The crickets spend their eight weeks of life listening to the local University of Oregon campus radio, enjoying the balmy environment of a climate-controlled facility, and munching down on non-GMO feed, organic fruits and vegetables, and spent grain from local craft breweries. When the time comes, they're placed in the freezer, fall into a hibernation-like state, and die.

Creating and maintaining Craft Crickets with his partner, Zoe, has been an incredibly important and rewarding process for Austin, but he hasn't always been so connected to his food. "It wasn't until later in life that I made the connection that every time I eat, I'm taking a stance on the environment and a true stance on my values." Food can be a form of activism and a daily personal practice in doing the right thing.

EATING BUGS FOR THE ENVIRONMENT

The crickets at Craft Crickets are 60 percent pure protein, and for every two pounds input, via cricket food, they get a pound of output, via crickets. The feed conversion ratios (the amount of food needed to produce a pound of body mass) for livestock are variable, impacted by things like environment, food quality, and whether the statistic is considering live or edible weight. Craft Crickets' 2:1 ratio is impressive: the feed conversion ratio for the edible weight of cows is commonly cited as being between 16:1 and 25:1, around 9.4:1 for pigs, and about 4.5:1 for chickens.

The modern industrial agriculture complex is one of the most destructive forces we level against our environment. "A lot of people don't see that connection [between food and the environment]," says Austin, "and through farming crickets we're able to help open up people's eyes between what they eat and what that does to the planet." While feedlots require extensive amounts of space and the concentration of animals can result in catastrophic pollution, crickets require only a small footprint. A pound of crickets, as opposed to a pound of beef, requires far less water, food, and energy. It represents less space, less pollution, and less stress on an already overtaxed system burdened with the colossal responsibility of producing more animal products than the world can reasonably sustain. For Austin, helping people see the relationship between their food choices and how the world will be used is an ideological background behind Craft Crickets. "That's almost the primary motivator, more so than we just think it's cool to get people to eat bugs. It's to come up with a nonconfrontational way to get people to think about their food choices." Food is not just food. It is a political statement that has intense ramifications for land allocation and the global economy.

I cannot wait to eat the crickets or to feed them to my chickens. The chitin in the crickets' exoskeletons can support my chickens' immune systems. Research published in the *Journal of Economic Entomology* has shown that supplementing chickens' conventional diets with black soldier fly meal didn't adversely affect growth rates or the flavor of their meat, but it did result in cheaper food costs. It might seem strange to be excited for a meal of insects, something so many try to keep out of their food, but I've already tackled dumpster food and figure I might as well continue on this strange trajectory of alternative eating. It's not for everyone, but it feels like a reasonable choice for me.

But first, they need to breed. "You've probably noticed that insects proliferate like crazy whether you try to breed them or not. They just do so well without consuming a lot of resources," said Austin. My wax worms bit the dust, but I have hope for the crickets. I clean out the plastic bin I used as a brooder for the chicks, cut a square out of the lid and hot-glue fabric over it for ventilation, and fill the bottom with a layer of vermiculite. Vermiculite is often used as a soil additive to increase moisture and improve aeration: it looks a little like perlite (those white pebbles you might find in indoor potting soil), but it's tan, finer, and spongy. I give them cardboard egg crates and empty toilet paper tubes to hide in. On a small piece of plastic cut from a jug, I drop cotton balls dipped in filtered water for sustenance, give them scraps from the kitchen, and provide a small shallow dish with dirt inside to lay their eggs in. I fix humidity and temperature gauges inside their bin and provide a red light for when the temperature drops below eighty degrees. I imagine it will only be a matter of weeks before I can start making cricket burgers.

"You don't need a lot of space or capital," Austin told me, and I think of him as I monitor the new roommates. The crickets are your average brown bugs: their legs have slightly serrated edges, their curious antennae wander around in front of them, and their bodies are barrel shaped. "It can be a hobby, just a couple minutes a day, and you can do it in a smaller footprint, arguably a much smaller footprint than chickens. You can have it in your closet," he assured me. I tell the chickens about my new project and promise them future treats.

As time goes on, it becomes apparent that maybe you don't need space or capital, but I must be missing something. The crickets seem to be doing well and then they slowly but surely die. I cannot figure out why, other than I am inept at keeping anything smaller than a baby chick. Not a single egg is laid, although I hear weak chirping once or twice. As a teenager I kept pet frogs and accidentally let the crickets get out of hand (read: they bred, escaped their enclosure, and then bred some more in the dark corners of my bedroom), so I am shocked that I cannot replicate the experiment given the nice food, comfortable bedding, egg crates, access to water, and clean substrate.

I order another batch online and get serious. I consider the possibilities: Did I monitor their humidity too closely? Was the temperature an issue? Did something mold? I come up with nothing and decide that if I have committed any sin in their care it was that of a helicopter parent, trying too hard to sustain a species that, if the cockroaches are any indication of the vitality of insects in this environment, should be able to self-regulate their own existence. I try again and commit to only giving them a few minutes of my day, as Austin suggested was possible

I mentally prepare myself not to fidget them—my third batch of insects—into nonexistence. Even though they have yet to arrive, I hopefully imagine what food I might create with their bodies. I contemplate mixing them with oregano and sprinkling them on melon and leftovers for the chickens and wonder how they would be in

CHAPTER 7

pasta. I am not the only one; people have become seemingly enraptured with the idea of fancy bugs as food.

Joseph Yoon, founder of Brooklyn Bugs, began incorporating bugs into his cuisine in 2017. His company promotes entomophagy awareness through creative culinary programming at conferences, festivals, and college campuses. His dishes are beautiful and his creations colorful: wheaty locusts suspended on charred asparagus spears; amber silkworm pupae on brown fried rice, sharp green broccoli, and fire-engine-red Os of sliced peppers; and cricket-crusted tempura shrimp, colored a shade between salmon and unfinished pine, atop a bed of celery and red-hued black ant gochujang sauce.

Even though Joseph refers to himself as an "edible insect ambassador," he wasn't always a fan of eating them. "When I first started," he admits, "it was hard for me to eat bugs on a regular basis." I think he is not alone and that most Americans would feel similarly challenged if they committed themselves to consuming insects. Joseph would eat bugs as a matter of research but not necessarily as a pleasure. Over time, things changed. As he worked with them, he learned to appreciate their flavors.

One of Joseph's goals is to make edible insect dishes that anyone will like. Sometimes, when he doesn't put the bugs center stage and uses them for garnish, people can be critical, but Joseph sees this as a wonderful application for roasted bugs. Sometimes he crushes them and mixes them into batter. If you're uncomfortable with bugs, why not start with some cricket-encrusted fried tofu balls or a salad with a soft mist of larvae? Jumping off the deep end and having full-blown, nothing-but-slug tacos seems like a recipe for disaster if the goal is to get people to eat more bugs. If food is activism, then entomophagy can help solve a global hunger crisis. Yes, the stakes are *that* high.

Even if people aren't willing to eat bugs, getting people more comfortable about bugs as a food source could be an important way to handle increasing caloric needs as the global population grows. Demand for chicken meat in the United States has been on the rise, and feeding broilers can require an immense input of resources. If some of the soybean and fish meal fed to broilers can be safely replaced with insects, the agricultural footprint represented by chicken meat can be reduced.

Of course, it'd be arguably more efficient if we just ate the bugs ourselves rather than feeding them to chickens. Joseph knows that "there's not going to be one silver bullet" that all of a sudden gets people to start eating insects on a meaningful scale, so he roots his approach to cooking them in "the fundamentals of gastronomy." He's constantly tasting his dishes and gauging what they need—more salt, more acid, more sweetness.

I ask Joseph how people who are just starting to experiment with bugs in their kitchens should begin their forays into entomophagy. "What if I'm not a trained chef?" I ask.

"First, you want to get a sense of its flavor," he responds, "and then you want to try adopting it into, I would suggest, your favorite dishes." He says that if you love mac and cheese, then try adding bugs to your mac and cheese. "I just find that it's so adaptable and the applications are really limitless," Joseph continues, before adding, "I would just say incorporate it into your favorite foods."

The replacement crickets arrive in the mail, and I immediately separate them into two factions: the freezer faction, so there will still be chicken treats even if their comrades do not survive the undeniably, yet inexplicably, harsh conditions I am offering them, and the habitat repopulation faction, which I intend to breed. I put the first sect in a breathable cardboard box with small holes and plenty of clean water so they can purge their system for almost two days, defecating whatever wares they ingested along their stint as USPS cargo. From there I put them into the freezer, where their body temperatures lower until they fall asleep and don't wake up.

The rest go into their new home, an improved cricket tank with a wire top to allow for better ventilation. They swarm the empty toilet paper rolls and collect along the dark insides so when you look through the tubes they appear tentacular with small moving antennae. They feast on processed cricket food, a capitalist luxury I allow out of sheer desperation, wondering if perhaps my prior rations of bread and kitchen scraps were nutritionally insufficient. As far as I can attribute emotion to insects, the crickets appear happy. They chirp at night, and after a week, I give them a small tray of moist soil. The females attack it with gusto.

You can tell a male cricket from a female by a long spike extending past their abdomen: males have two, while females are furnished with an extra protuberance. This is their ovipositor, an organ meant to inject eggs into appropriate terrain, allowing them to hatch below the surface. They jab their ovipositors into the black dirt, waiting with their backs hunched. I watch them as they hump the soil in reverse, trying to dig a good hole for egg laying, and then as they sit there poised. Some crawl onto the others who are laying while they remain immobilized. The bump and grind that follows reminds me of middle school dances.

Some crickets miss the soil and lay their eggs in the nearby vermiculite. All seems to be going well. Within an hour, the anarchy is over, their biological feats completed, and as if the chaperones turned on the gymnasium lights, the crickets scurry back with the rest of their classmates to the nearby buffet tables. When I remove the soil, small translucent eggs like miniature grains of rice are scattered within it. I take a handful of the vermiculite around where the highest concentration of crickets dug their protrusions and place it with the soil into a secondary glass container. I put a thin cotton cloth over the top, spray it with water to keep the soil most, and leave it in a warm part of the house along with the cricket tank. I have to wait four to seven days to see if it worked, and over the course of this time I spend my few daily minutes making sure the soil stays damp and waiting for pinhead crickets to emerge.

About two days after the frenzy of egg laying, blight appears to sweep through the cricket population. They die at staggering rates, and although I cannot be sure, it appears I have lost 50 percent of faction two. Bodies are everywhere in a kind of mass extinction. I do not know what the problem is, only that it smells. I remove the egg crate with much shuffling and several insect escapes, regrettably transferring some of the remaining live crickets from the safety of their tank into the wilds of the living room, where I'm sure they'll, at best, die in dark cracks or behind bookshelves and, at worst, provide fodder for the cockroach hoards that constantly threaten to establish a foothold. I remove the dead bodies, replace the egg crate with more stockpiled toilet paper tubes and promise them that I will disinfect the tank and change the vermiculite in the morning.

The next day when I wake up, only ten crickets remain in the large terrarium. I know because I count them. The rest are dead, bodies upturned and blackened along the bottom of the cage.

Cherishing my surviving crickets, I think back to something Austin told me. He said, "It can be a hobby, just a couple minutes a day." Clearly, I am not suited to this type of farming. The overworked

CHAPTER 7

method failed, the relatively hands-off method failed. My population is down to 1.4 percent of its original size.

A week later, just as I was hoping to see small crickets hatch from their translucent eggs, I find the final seven crickets dead and the babies fail to emerge. Resisting the inevitable, I keep the soil moist. Around day fourteen, I throw the dirt and unborn crickets into the garden.

At least, I tell myself, there are crickets in the freezer at the end of this wasted escapade. I pull the box of crickets from the freezer and tap them out into a colander, a step I deemed necessary as soon as I peered into their cardboard coffin. Small flakes line the bottom—feces, I suspect, even though I'd withheld food before transferring them. After a few shakes of the colander I transfer their frozen bodies to an ungreased cookie sheet. I resist the urge to wash them, because, as I'm told, this rinses off some of the flavor. During the shuffle a number of their legs fall off while their attached appendages stick out like those of unbalanced dancers. I slide the metal sheet into the oven, preheated to 200 degrees.

After an hour, I test one. They're supposed to be dried out and crunchy, but the initial sampling reveals a slight dampness inside. They taste nutty, vaguely salty, and bread-like. I decide to let them roast for another twenty minutes. During this interval I go outside and notify the chickens.

"Will you eat crickets?" I ask. Amelia, seeing I haven't brought her anything, pecks my leg. I can never tell if it's frustration or hopefulness, but the girl knows how to bite. I flinch and she backs off before trying a strand of my hair. Joan watches. She's so easy-going I imagine she'd eat any kind of dried bug.

When the crickets are finished, I don't immediately go out to the coop: rather, I do what Joseph suggested. I experiment.

I crunch down on one cricket, then another. Nutty, still. Savory. They remind me of something wet—they taste like that damp smell, but loamier and in a surprisingly good way. In a jar I mix chili powder, cayenne, and salt to flavor them. The spices don't stick very well, so I add a little oil. Although they're unconventional, they're pretty good. I crunch a couple more.

The chickens are a harder sell. Amelia pecks one apart without seeming to consume any portion of it; her attack is investigatory rather than gustatory. Joan, as I expected, is elated. She snaps them into her beak before looking for more. When I fail to produce any immediately, she stretches her neck and eyes my face.

I pop one into my mouth and offer another to Joan. Amelia cackles and yells.

"Eating bugs will save the world," I tell Amelia, trying to reason. She pecks my shirt, expecting something better. I don't know what her problem is, but she's clearly my chicken.

Joan looks at me quizzically. "More?"

8

Productive Pets

THE RISE OF BROILERS

In the morning when I hear Katie emerge from her bedroom, I run to head her off in the hallway and try to look calm. I'm wearing the boxers and the tank top I slept in, and I have pine shavings in my tangled hair, a morning aesthetic caused by cleaning out the brand-new five-by-eight-foot pallet-and-plywood brooder. It's a recent addition to the sunroom. I block Katie, knowing she is not a morning person and just wants to get a Diet Coke from the fridge on the back patio, and blurt out, "There's something going on outside." I know she doesn't prefer to start the day with the sight of blood.

This is not how I envisioned my morning going either.

"There's a chick tied to a chair bleeding out into a plastic bag," I announce. "I'm sorry." The chicken was just shy of two weeks old, dead a full six weeks before I planned on slaughtering her.

The new chicks—eight of them in total—arrived in the mail when they were just a couple of days old. They are piranha menaces who flock to the feeder as soon as I place it in their brooder, jump up at the dangling ends of my string bracelet, and often successfully land a bite, only releasing when airborne. They crawl all over each other with fat, squatty legs, climb atop the backs of their sleeping siblings, and peck each other's sides. They are all bottom-heavy and small-winged and chirping constantly. They guzzle water, then fall asleep against the metal rim of the trough. They are adorable and feisty and not what I expected. After only a few days, it became clear that I wouldn't be able to accommodate their rapid growth if I kept them in the bathroom. It would only be a matter of time before they were not feathered enough to live outside without heat lamps but too large to remain in a brooder under the sink. I drove around Lafayette picking up pallets and plywood from the side of the road, reused some scrap from another project, and sawed apart the wooden remainder of Katie's old bedframe when she bought a new metal one. It took me a week to gather materials and a single morning to build the brooder, which is an overly expansive forty-square-foot affair with a hanging food trough and multiple heat lamps. They don't need all this space and most chicks don't get it, but I'm worried they won't get enough exercise without it.

The new chicks are not egg-laying hens. They are Cornish crosses, also known as broilers or meat birds. They do not develop like regular chickens and have an incredible ability to rapidly convert calories into body mass in spite of their best interests. If I don't withhold food for at least twelve hours a day, their breasts and thick thighs can grow so fast the skeletal structure of the developing legs is compromised. Their legs are set wider than Joan and Amelia's were when they were chicks, and although they look like small tanks, they are fragile creatures, the fraught results of decades of selective breeding.

Cornish crosses are not your normal chickens. Humans have literally bred them for one purpose and they forever suffer the consequences. Even without overfeeding, Cornish crosses' massive girth tends to outmatch the ability of their bones and organs to support the rest of them. Their legs can fracture, hearts can fail, ligaments can tear, joints can disintegrate, and feet can blossom with sores from chronic pressure. When they reach maturity in commercial farms, more than 25 percent of broilers suffer from joint issues. This is why a large brooder is so important, with water and food spaced in opposite corners, the heat lamp in another. I've heard that with proper exercise they will do normal chicken things: forage, dust bathe, jump into coops, and crawl up planks. While they are inside, I make sure to give them space.

Cornish crosses have white feathers, red combs, widely spaced stout legs, and visibly protruding breasts. They are thick and don't stand sleek and upright like normal chickens. Anywhere from 10 to 60 percent of a commercial Cornish cross flock can become affected with ascites syndrome, or waterbelly, a condition caused by liver damage or the body's inability to overcome the force it takes to pump blood, which results in accumulation of noninflammatory fluid in abdominal cavities or organs. The poultry industry raises almost nine billion of these Frankenchicks every year: in factories they often struggle to stand under their massive girth and wallow in their own feces when moving becomes too uncomfortable or tiresome.

It's clear that Cornish crosses, like most things in modern agriculture, were carefully engineered to produce the greatest yield with minimal input, regardless of the consequences. Geneticists, scientists, and hobbyists all worked to sacrifice hardiness and normal organ, muscular, and skeletal growth for the sake of thick breasts, tender meat, and lower caloric requirements—things that consumers care more about than blown ligaments and joint pain. The process of creating Cornish crosses began in 1927, when Oliver Hubbard and A. W. Richardson started using selective breeding to concoct a new kind of chicken. It took them seven years to complete their prodigy, the New Hampshire. These selectively bred descendants of Rhode Island reds grew faster on less food than typical egg-laying hens. They could be slaughtered at younger ages, meaning farmers could invest less time for greater yields. Additionally, the meat was softer and more tender since the muscles of young chickens aren't completely developed through age and exercise. In the seven years that followed, the percentage of broilers bred in the Northeast doubled from 40 to 80 percent, and a new era of selectively breeding chickens for meat began.

While the starting gun might have gone off in 1927 with Hubbard and Richardson, the race didn't get underway until 1946 when the Great Atlantic and Pacific Tea Company (A&P) announced its Chicken of Tomorrow contest: a monetized challenge to develop a fast-growing, calorie-efficient, big-breasted broiler chicken.

It is worth remembering that these were changing times, as far as our national relationship to chickens and chicken meat was concerned. It was in 1923 that Cecile Long Steele launched the first ever American chicken farm due to a lucky clerical error and her own impressive entrepreneurship, and by 1926 there was enough of a market that she'd financed the construction of a ten-thousand-chicken henhouse. Until 1940 most of the chicken meat consumed in America was from backyards, and the per capita consumption of chicken was around ten pounds annually. The first year since 1909, the beginning of the USDA's Food Availability Spreadsheet Recording, that annual chicken consumption reached anywhere above 11.1 pounds was 1942, when the pounds per person jumped to

12.7. The year after, in 1943, Americans ate, on average, 16 pounds of chicken.

It's likely A&P saw opportunity in these cultural and culinary shifts. Broilers, which could be cut and prepackaged, fit the self-service model of A&P supermarkets. Chickens could be dropped off in advance and housewives could pick up easy-to-cook meat, choosing between thighs, breasts, or whole roasters. There were no custom cuts, no fuss, just ease and efficiency. In "The Surprising Origin of Chicken as a Dietary Staple," Maryn McKenna notes that Howard C. Pierce, the poultry research director for A&P, envisioned "a chicken with breast meat so thick you can carve it into steaks, with drumsticks that contain a minimum of bone buried in layers of juicy dark meat, all costing less instead of more." The winner would be given $5,000, equivalent to nearly $55,000 in 2020. To the benefit of consumers and the eventual, unfathomable detriment of commercial chickens, the contest was an incredible success.

McKenna described A&P's contest as "massive" and a "significant challenge," and it was undeniably so. "The contest had 55 national organizers—scientists and bureaucrats loaned from government agencies, producer organizers, and land-grant colleges—and hundreds of volunteers in 44 states," she writes, noting that since Alaska and Hawaii had not yet been admitted, this number was out of 48 states at the time.

Farmers and breeders from across the country entered the contest, submitting eggs to specially built hatchery facilities where chicks were raised in controlled environments and fed a uniform diet. At twelve weeks old the chicks were slaughtered, weighed, and judged for their edible meat. Following a series of state and regional contests, forty contestants were chosen to compete for the national title. In the end Charles Vantress of Live Oak, California, won by crossing a Vantress red—his own take on wide-legged, big-breasted Cornish chickens—with a New Hampshire, the broiler chicken developed by Hubbard and Richardson in the late 1920s.

The tournament turned out to be as forward thinking as its name assumed, and today almost all chickens raised for meat in the United States are in some way born from its results. While farmers had been historically leery of crossbred chickens, worried their young might be susceptible to genetic issues, the A&P contest changed everything. Vantress won with a crossbred bird in 1948 and did it again when the contest was repeated in 1951, thereby proving their worth to the poultry industry. McKenna writes, "The winners of the Chicken of Tomorrow contest did more than create new birds; when they transformed chickens, they recreated the chicken industry too." By 1959 Vantress's hybrids sired 60 percent of broiler chickens in the United States, totally displacing the variety of purebred chickens that were once kept as egg layers in homesteads and backyards and eventually consumed as spent hens. It was a revolution much like monocropping: diverse chicken breeds of the past, specific to different locations and purposes, were replaced with one meaty bird.

While the original goal of the Chicken of Tomorrow contest was to sell more chicken to American consumers by creating birds that were more palatable and reached market weight sooner (read: cheaper to produce and purchase), a side effect of the genetic triathlon was that breeders became more comfortable with crossbred chickens. The broiler industry skyrocketed, and most famers retired from tinkering with genetics to find the perfect chicken (it had already been discovered), while industry breeders continued their projects of creating complex crosses prized solely for their ability to grow larger than the last iteration. These hybridized birds were amalgamations of both desirable and undesirable traits, requiring careful industry

CHAPTER 8

record keeping and the continual renewal of specific breeding combinations. Famers couldn't breed their own chicks from the roosters and hens they bought from breeders because the offspring of hybrid chickens would express a variety of traits, essentially performing a kind of chromosomal self-destruction within a generation. This is the same reason farmers don't save hybridized seeds today: it just won't work as well as buying a new batch.

Commercial Cornish cross chickens reach slaughter weight in six weeks or less, although the ones I raise in my backyard will take eight to nine weeks to fully develop. The male chicks in my brood will weigh approximately ten pounds and the females approximately eight, but in commercial farms they can reach twelve pounds. The differences in weight and age are in part environmental: my chicks will experience normal light periods rather than artificially extended ones that promote rapid growth, and they will not consume antibiotics in their feed. Although slaughtering something that is only two months old might seem inconceivable, it is not generally recommended or considered kind to keep them much longer. Even if someone wanted to keep them as pets or companions, Cornish crosses will struggle to survive, making their life expectancy comparable to—or worse than—your average hamster's (which is eighteen to twenty-four months).

According to Isabelle of Clorofil micro-sanctuary, broilers are "really hard to keep alive for more than a year." When I visited the sanctuary, Isabelle's husband, Peter, said, "Nobody ever sees broiler chickens because they're rarely in sanctuaries because they die all the time." Even rescued Cornish crosses cannot escape their broken genetic code. "Their skeletons are collapsing under their weight, they have heart failure, and," said Isabelle, "they're only six weeks old. Six weeks!"

PRODUCTIVE PETS

The chick I accidentally killed seemingly died from a brain injury when she collided with my hand. I was changing the shavings and the chicks were scuttling around, excited to see my hand, which always translates to food. Her small head and swollen body bounced off one of my knuckles as I plunged my hand into the brooder to scatter clean bedding as she simultaneously charged.

"Oh buddy," I crooned. "Are you okay? That must've hurt."

The chick looked up at me, shook it off, and walked to the other side of the brooder. When I glanced back, the chick was on her side convulsing. One of her legs was drawn toward her body, the other kicking. Her head twitched. The other chicks ignored her.

I picked her up and wondered what to do, or more accurately, when to do it. I knew she wasn't going to get better, but I didn't know how long to hold her before getting a knife. Perhaps I should've just done it instantly, but I squatted on the concrete floor in shock. Within twenty seconds she stopped moving. I put her limp body on the windowsill, went back to spreading pine shavings, came to my senses, thought of how blood might pool, and ran to the kitchen for the knife so I could bleed her out and salvage the body. The blood ran slowly without a heartbeat to pump it.

"I'm sorry," I repeat to Katie, still blocking her way. Later I would wonder whom I was addressing and to what I was referring. Was I sorry for making Katie see blood first thing out of bed, sorry for buying Cornish crosses, or sorry for the ways I'd contributed to and benefited from the fragile genetic code of a bird born from consumerism's tendency to value money and instant gratification over animal welfare, ethics, and compassion?

CHAPTER 8

She told me it was all right, went outside to the outdoor refrigerator, and came back in seemingly unfazed. The death toll in this house is rising: two laying chickens, two dozen wax worms, a staggering number of crickets, and now a small chick. I was sorry for all of it, but also not: any apology would be too tempered by curiosity to be pure.

How can we not be curious when we know what we've been told about an industry as massive and far-reaching as animal agriculture is full of misconceptions? The chickens most people eat on a weekly basis are not really chickens at all, at least not in the way we normally think of them. Their bodies grow so rapidly that they walk on their lower legs in the poultry equivalent of an army crawl. They are so fragile and their organs are so stressed that small collisions can prove fatal. When you lift them there is not light resistance under their downy feathers: their bodies are soft putty, breasts warm, chests malleable like water balloons. If I am trying to understand chickens and the poultry industry, how better to do it?

Finally, their productivity is undeniable. They convert calories to body mass far more efficiently than other chickens, which means they're particularly economical contributions to homesteads and small farms. If people want to raise their own chicken meat, Cornish crosses are far cheaper and faster than other options. Even though there are elements of irrationality—their health risks and the strange way they rapidly acquire mass—involved with Cornish crosses, they are undeniably efficient to raise for meat.

These small wonders of selective breeding both delight and terrify me. At two and a half weeks old they chase each other, charge with wings clapping, ravage the feeder in a frenzy of beaks and scratching feet, and eventually fall asleep in cuddle puddles, their necks stretched over the immense soft backsides of their brothers and sisters. I watch them for hours, worried about their strange genetic codes and in awe of how cute they are.

There's a strong tendency toward "them." I dote over the chicks, clean their water, swab their vents with damp Q-tips, and all the while use collective pronouns. When Grendel, Francis, Joan, and Amelia were in the bathroom brooder, I could differentiate between them early and named them within a few days. With the nearly identical Cornish crosses, keeping track of who is who seems almost impossible. They all look the same, besides minor facial and size differences that only appear in moments of scrutiny and direct comparison. This buffers my attachment, but not my commitment.

One night, Katie and I find ourselves squatting awkwardly by the edge of the brooder. One of the chicks seems to be getting picked on—another chick pushes him away from the feeder, and then another chick crawls under him and sends him toppling. He is often in the corner alone. We fall in love and name him Jet.

This is not an outrageous tendency or an isolated phenomenon. An ethnographic analysis of the relationship between farmers and their livestock uncovered that human-animal relationships are negotiated by degrees of attachment and detachment. The four different attachment styles were identified as concerned detachment, when animals weren't considered as individuals but taken care of as sentient creatures; detached detachment, which is the kind of attitude exemplified by poultry farms that just consider chickens as pure commodities; concerned attachment, or the sense that animals are individual, decommodified, and experience a degree of meaningful interaction with their human caretakers; and attached attachment, which is when animals are individually recognized and experience an even greater degree of positive, meaningful interactions with the humans who care for them. What I feel for Jet, the special chick, the

CHAPTER 8

one who seems to be the perfectly human underdog in which I can recognize myself as a vulnerable mainstay against the animal hoards that seek to trample me, screw me out of resources, and push me aside, is somewhere between the final two categories. Rhoda Wilkie, the author of the study and a sociology professor specializing in multispecies scholarship, noted that it is common for farmers to develop soft feelings for a favorite animal. While I know this will not necessarily serve me in the slaughtering process and that it is a cheap coping mechanism that helps me hold the realities of the chickens' situation at a remove, I cannot help it. Katie and I mark the tip of Jet's wing with a Sharpie and hand-feed him.

Before the chicks are removed from the brooder, I take a trip cross-country to California. I have the flexibility to take the time off, and I'm notorious for traveling home to see my parents whenever I can. I build a bed platform in my minivan, so I can sleep inside the car. I figure that heading into a tent every night with two dogs and two chickens would be difficult, plus potentially dangerous and not very stealthy. Although the thought crosses my mind, it does not seem possible to take the Cornish crosses with me, but there's no question that Tashi, Atlas, Joan, and Amelia are along for the ride. I plan to camp on Bureau of Land Management land and at free, rural campsites. Katie agrees to take care of the chicks in the brooder. I set up a layered system for the bedding with garbage bags so it's possible to remove a layer of plastic from the bottom up, shavings and all, and toss it out. I make sure they're stocked up on food and that the waterer is secure. She assures me it'll be fine and sends me pictures of the chicks to confirm.

While I'm on the road, I let the chickens hang out in a pop-up pen meant for small dogs at rest stops, gas stations, curbside pull-outs, and campsites.

Camping with chickens might seem strange, but it's a surprisingly straightforward process, and Joan and Amelia fall into the rhythm that the dogs set. After a few hours of driving, when the dogs need to stretch their legs or use the bathroom, the chickens get a break in their fenced area. People stare and point while I sit beside the van parked in pullouts and scenic overlooks, the dogs off-leash and gnawing sticks, the chickens dust bathing and snacking on whatever local cuisine that patch of ground has to offer. The entire process is as outrageous and decadent as it sounds.

On the trip I think about how things that seem impossible are sometimes so easy. The car doesn't smell like feces, the dogs don't bother the chickens en route, and there's plenty of open space to camp on to keep everyone happy. It almost seems to go too smoothly, and I begin to wonder about who gets to make the rules. Who says you can't camp with chickens? Who says they can't travel with you? Sure, I could have left Joan and Amelia in Katie's care, but I didn't want her to have to deal with chickens both inside and outside the house. Plus, I begin imagining the rocks from different states accumulating in their crop. Chickens will eat small rocks and grit to help with digestion, so I picture them nabbing stones from our various stops: small pebbles from a public park in New Mexico, desert dust from Arizona, granite from Sequoia National Forest.

On the return leg, I wake up in Texas cuddling my dogs. It's cold enough to wear socks and a beanie to bed. From where the dogs and I are lying on the bed platform with the chickens at our feet, against the back hatch of the minivan, I can hear Amelia making soft noises. Amelia, an early riser, is talkative. Although she isn't one to generally sing an egg song, which is a common name for the vocalizations that chickens make after they lay an egg, she will chatter quietly when she wakes up (the word *song* here is misleading, for this series of *bawk*

CHAPTER 8

bawk bawk backawwk noises is not anything like the melodic tunes that songbirds create).

While the sun comes up over the red Texas rocks outside the van, I listen to Amelia begin to wake up. Tashi is still snoring. I think about how the five of us constitute a weird mobile chicken coop. I wonder how long I could sustain this kind of travel.

Around midday while I'm stopped for lunch, Katie calls. "The chicks are out of food," she says. I am flabbergasted. When I left, they had what I thought was an incredible surplus, not counting the gallon of food I'd dumped into the feeder. Although I planned to spend another night on the road, I decide in an instant to drive straight home with my animal hoard, partially because I'm only twelve hours from home and partially because Katie broke her thumb falling out of the enormous brooder while cleaning it.

I arrive in the middle of the night, and when I wake up in the morning, there's anarchy. Jet is perched atop the beam that supports the hanging waterer. The chicks are four weeks old, and neither Katie nor I is surprised that our favorite chick is the one to create chaos. He looks at me expectantly while sitting on top of the beam, ignoring my questions about how he scaled the seemingly impossible impasse of thirty-six careful inches of plywood. I gently grab him, place him back on the pine shavings, and request that he cease and desist his aviation experiments, at least until I've had some caffeine to counteract my late-night arrival and early-morning trip to the feed store.

Soon after, as I'm eating breakfast and drinking coffee on the back patio, I glance through the window into the sunroom to see two small heads through the window.

Jet plus a companion chick are back up on the waterer. I go inside, put them on the pine shavings, and make more coffee. Katie joins me outside for a caffeine fix, and almost immediately, we notice someone watching us from the windowsill. There's a small beak pressed against the glass.

"It's Jet," she says, taking a sip from her mug. "I'm glad you're home."

"Me, too," I tell her. She's the only person I would trust with the chicks, although it's clear they are on the verge of a revolt.

We talk about how there's no way they can stay inside the brooder. At four weeks old they are fully feathered and able to withstand the elements. "I just hope Amelia doesn't try to kill anyone," I tell Katie, and resolve to move them as soon as the coffee's gone.

While chickens might seem harmless, they are not. *Pecking order* is a literal term: hens and roosters will harm each other to establish dominance. Aggression decreases once dominance is established, and although any sustained injuries are usually minor, I've broken apart chicken fights that left one or more hens bloody. Sometimes, a chicken will go for the comb and waddle of another to negotiate her social position. If the chicks are going to be indoctrinated by violence, at least let them be hefty first, I figured. While Joan has been known to chase squirrels with a vengeance that can only be interpreted as bloodlust, Amelia can be downright apocalyptic; a pterodactyl swarm of feathers and claws, she's successfully chased Atlas away from the chicken run fence, chortling war cries. When she "mistakes" the freckles on my arm for edible matter, she leaves marks.

In 1921 Norwegian zoologist and comparative psychologist Thorleif Schjelderup-Ebbe published a dissertation on the social behaviors of chickens, partially inspired by a childhood obsession. When he was six, he became preoccupied with chickens and began identifying and naming his grandmother's flock. At ten he started keeping notes about the chickens' hierarchy and rotations of social position. He coined the term *pecking order* to describe the ways chickens maintain their social hierarchy. The head chicken—or, as Schjelderup-Ebbe referred to it, the *despot*—gains access to food before other hens and pecks subordinate chickens without fear of retaliation.

In chicken societies roosters are dominant over hens, a feature Schjelderup-Ebbe thought helped ensure procreation. Studies show that chicks don't exhibit competitive behavior until three days old and don't begin fighting to establish their pecking order until they're sixteen days old, but in all-female flocks baby chicks might not establish a social hierarchy until seventy days. While no one knows exactly what makes one chicken the boss and another subordinate, it doesn't come down to how big the chickens are. It seems to have to do with their personalities. Aggressive dispositions feed into the making of a despot, but so do social connections: higher-ranking birds make more contact with others than lower-ranking ones. Leadership behaviors, like leading other chicks to a heat source, can also be used to predict which chickens might be in charge later. When Amelia learns how to hop on my arm after only a few repetitions and Joan doesn't get it for weeks, I am not terribly surprised. We all know who the boss is.

There's no getting around it: impending bloodshed or not, the Cornish crosses need to be introduced to Joan and Amelia. They must be moved to the outdoor coop.

I bring the chicks outside and set them loose in the run. At first they stay clumped together, all pecking the same patch of dirt. Amelia choruses, walks over, and promptly pulls out a feather from someone's neck. She ruffles her own feathers and puffs up her breast. She walks from chick to chick pecking them on the neck or head and is particularly rough on the young roosters. After she's satisfied that everyone knows that she is in charge, she settles into a dust bath with someone's feather still stuck to her beak: mission accomplished. Joan watches from a distance. Despite their differences in age, the chicks are almost as broad as Amelia. By the end of the day they get along smoothly—Joan avoids the newbies, and the newbies, understanding their place, avoid Amelia—so I put them in the coop for their first night outside.

In the yard, the chicks' personalities become dynamic. They charge the fence when they see me coming, running on their thick, wide-spaced legs like sumo wrestlers while flapping their wings for increased velocity. I must tread carefully through the pen or risk stepping on one; they swirl around my feet like cats hell-bent on broken hips. I begin to fall in love with them and feel like they trust me: they seem to enjoy being petted and swarm when I crouch down.

Joan and Amelia are not as enraptured with the chicks as I am. Joan likes to charge them during their moments of relaxation, sending the young flock into a flight-happy frenzy. Amelia periodically lays into a chicken if it tries to stand next to her at the feeder. She seems offended and confused by their presence. "They'll be gone in a month," I confide to Joan and Amelia. "You won't need to deal with this much longer." I begin serving Joan and Amelia their morning meal in a separate container at the far end of the run, mixing their pellets with dehydrated mealworms and water. The other chicks don't dare sample from their delicious breakfast. Pecking and ripping out feathers might be a brutal way of establishing social control, but it works.

While the older chickens might seem brutal, if the worst thing that happens to the broilers on any given day is Amelia's vengeful beak, they're admittedly doing pretty well. Chickens bred for meat are raised in sheds the size of football fields where they breathe air polluted with ammonia, dust, bacteria, and fungal spores. The stocking density—the amount of birds per unit of floor space—for commercial chickens is based on weight, not number of birds. The National Chicken Council's recommendation of 7.5 pounds of live bird per square foot of floor space translates to approximately a bird and a half per square foot. Under these conditions, chickens develop lesions and sores and gouge each other's backs as they climb over neighboring birds.

According to the *Commercial Chicken Production Manual*, "Limiting the floor space gives poorer results on a bird basis, yet the question has always been and continues to be: What is the least amount of floor space necessary per bird to produce the greatest return on investment?" On a "bird basis" chickens might not have room to flap their wings, nest, or roost; in other words, they can't do anything that makes them birds. Chickens are given four hours of darkness per day, but the intervals may come in one- or two-hour increments. This artificial system sacrifices their circadian rhythms and healthy bone development for the sake of continued appetite.

Before the chickens suffer transportation to slaughterhouses, they must be caught and removed from the henhouse. During this process chickens are carried upside-down by their legs and tossed into transport cages, a painful process that harms their already brutalized bodies, resulting in bruises, dislocations, and broken bones. A worker might be expected to catch *and* transfer as many as fifteen hundred birds per hour (that's twenty-five birds a minute, or a bird and a quarter every three seconds). There are machines that can accomplish this process, which sometimes use spinning rubber fingers to grab chickens and conveyor belts to transport them into cages. These automated methods do save workers from doing this job, but they can also be traumatic for chickens.

Transportation to slaughterhouses is no less traumatic. Chickens freeze to death in the winter and die of heat stress in the summer. According to the Humane Society of the United States, of the 8.5 billion broilers that are brought to slaughterhouses annually, somewhere between 10 and 39 million die on the way. The most common causes of death are ruptured livers, ruptured lungs, head trauma, and asphyxiation, in that order. The cost of these deaths is not as astronomical as you might imagine: a poultry processing plant that slaughters 1.25 million birds a week might only lose $325,000 on dead on arrival (DOA) birds. When a study published in the *Journal of Applied Poultry Research* says, "the problem has received little research attention in the U.S.," it is not a conundrum as to why. DOA chickens are a cheap cost of doing business. They're a side effect of rough handling, lack of environmental controls in transport trucks, and overcrowded conditions. John Webster, professor emeritus at the University of Bristol, describes the broiler industry as the single greatest example of humans' inhumanity toward another animal. It seems unlikely that industries will change until people make them.

It is hard work trying to make changes. It is hard work to raise chickens for meat and hard work to feed these extra hens. The chicks have become voracious and can no longer properly be called chicks. They have grown so large that they are not flying or running as much as they used to. Sometimes, if they only want to close a short distance to the feeder or find a new place to dust bathe, they walk like soldiers, army-crawling on their whole legs instead of just the pads of their feet in a crippled-looking crouch. It is hard to watch them and feel only anticipation about eating them or endearment over their clumsy ways. There is both nervous hunger and care.

They still fly from the coop in the mornings when I open the doors, but I've started encouraging them not to, worried they'll crush

CHAPTER 8

their legs as they land in a frantic flurry of wings, their descent more and more tantamount to a free fall with each passing day. Instead, I place them gently on the ground. Some of the larger ones have even come to expect this and wait perched on the coop's wooden lip.

The work sometimes feels ridiculous. It seems like a lot to take on and there's always that looming question: Why bother? Why not just buy chicken from a store? Why go through all the trouble of raising them, not to mention the toil of killing and cleaning these animals that I care for? The answers don't always come easy, but sometimes they do. It is hard work to accommodate them. I worry about their hearts and legs as they grow. And I worry about my own heart in all of this, but I would not choose to invest myself in another project. There are no simple ways out, and sometimes working for the things you want is the best and only answer.

9

Slaughterhouse in the Backyard

CULLING HENS

WHEN I GET home, a woman I've met only twice before is in my backyard wearing overalls and sensible shoes. "I am so excited," Delaney beams. I step out of the car and sling my backpack over my shoulder; juggle my coffee cup, jacket, and house keys; and try to look friendly. I'm running late from a poetry talk, completely starving, and she's early to the chicken workshop. This afternoon I'm going to teach a small group of people how to slaughter hens. None of them have seen anything like this before.

I unlock the back door and the dogs pour out. "I have been wanting to do this for a long time," she says, gesturing toward the chickens in the backyard as they run and scratch along the fence, "but I didn't know how to get in touch with it."

"I'm glad you're here," I tell her. "I'm really glad you could come." In reality, I don't mean it. I'm trying to be nice, but I feel petty and defensive and shitty and protective. I started a group text with the workshop crew a few days prior, hoping it would be a low-impact way

CHAPTER 9

to let everyone know who was coming and to disseminate important information like my address and a warning against wearing nice clothes. I expected everyone to be somewhat reserved, or at least considerate, but Delaney kept saying things that struck me as cavalier.

Maybe she's not insensitive; it's possible I'm the one with the problem. Earlier that day I fed the chickens a small amount of grain and apologized. "I'm sorry you can't have a full meal," I told them. They swarmed my feet, asking for more. One of the roosters tested a chunk of my palm, biting down hard on the fleshy pad with his beak, and twisting. "Hey!" I jolted. He looked at me, perplexed. The others pecked at my back, my ring, and my hair, their want intense.

I wasn't withholding food to be cruel. I was being practical. I didn't want their crops, the organ that holds their food before it goes to their gizzard, to be full. Puncturing a crop when you eviscerate a chicken is easy and messy: the elastic organ's thin membrane is resistant to tearing when you pull but like butter under a sharp knife. If you cut into it, the grain, grass, and whatever the chicken dined on earlier that day spills forward, smelling a little like a combination of wet dog, sink bacteria, and the soft yeast of rising bread. After I yelled that morning, the chickens' spirits undeterred, they continued pummeling my body with their beaks, pecking the hair along my arms and legs, regarding me with sweetness and frustration. *Human, why are you not doing your job? Human, help. Human, I'm hungry and you're the source of all that's food and good.* The rooster who bit me shuffled alongside me, cocked his head, apparently considering the edible content of my left eyeball, decided against it, and let me pet him. The one day I don't hold up my end of the bargain—to provide for them, to look after them, to make sure they're content and physically comfortable—is the day I've come to collect their debt.

If you're a bird living on planet Earth, there's a good chance you'll find yourself in this position. A 2018 study exposed that only about 30 percent of Earth's birds remain wild; the remaining 70 percent are chickens and other domestic poultry. During a 2018 interview, Dr. Ron Milo of the Weizmann Institute of Science, who led the research, stated, "I would hope this gives people a perspective on the very dominant role that humanity now plays on Earth." We have become masters of our garden and are turning it into a factory farm. Mammals are suffering a similar trajectory: while humans make up 36 percent of all mammal life, wild animals comprise 4 percent, and the remaining 60 percent are livestock. The birds who live in the oak and fig trees nearby, who chorus from the lemon and orange trees in the yard, represent a small, wild fraction of Earth's avian life. The chickens that scrabbled at my feet for food that doesn't come—those overgrown children with their hefty breasts and predisposition to leg fractures and high risk of heart attacks—are the real representations of birds today.

Sitting with the chickens during their last morning, I thought about the workshop, what was going to happen, how charming and trusting the chickens were, how taken care of they were, and some of the things Delaney said.

I know this is what these chickens are for. I know this is why they have been bred to grow fast and fat and that their chances of living a "normal" life, one in which they would not die young from organ failure, was undermined from the beginning by their broken genetic codes. I know this is the situation of most birds today and that the care I provided for them far exceeded what they would've received on a factory farm. I know I raised them for meat and remembered, almost every day, that their blood would be spilled, but I don't like saying it.

CHAPTER 9

Dr. Gail Melson of Purdue University understands the dynamic pull of animals as both pets and livestock. In her research, she studied children's experiences raising animals as part of a 4-H project. The kids raised animals (usually pigs, cows, and sheep) and then brought them to auction where they were sold to the highest bidder. The money they earned helped their families and future and taught them about the economics of raising livestock. When I asked Melson about what she discovered, she said, "What I found was a complicated story, and definitely an attachment that would grow and a treatment of the animal as a pet, but at the same time there would be unconscious mechanisms that would help them [the children] to deal with the end of that story, which is the animal is generally auctioned off and goes to slaughter." One of the main unconscious mechanisms involved not naming the animals. "In fact," she said, "some of these children would talk about how their older brothers or sisters or friends would . . . [say things like] 'Don't give the animal a name because that's going to make it harder for you.'" While chickens in backyards usually aren't raised for economic reasons and moved into the category of companion animals, "in the 4-H study I found these two categories were mixed," Melson said. "The children were developing attachments. . . . There's sort of an inherent—maybe you'd say—contradiction in that you'd lavish loving care in order to ultimately kill the animal."

One of the ways we form attachments to animals is by grooming them. "For children, and maybe adults, too, that physical contact, brushing, getting the coat of an animal soft and sleek, that's sort of deep in our DNA in terms of bonding. Primates groom each other as a way of bonding, so I think there are processes that go on below the conscious level that promote a kind of attachment when you're caring for the welfare of another living being." When she said this,

I thought back to cleaning the Cornish crosses' vents when they were chicks, about petting them in the backyard while they gorged themselves on breakfast, and about how, only a few days before the workshop, I went outside with a couple of rags and a bucket of warm, soapy water to clean feces from their feathers and mud from their legs. Perhaps, accidentally, through the natural rhythms of care, I've become too attached. As they got older, I found myself spending more time with them, talking for them as if they were miniature people, and apologizing to them when Amelia picked fights. "She's just feisty," I'd say, extending a chunk of bruised, dumpster-salvaged banana their way to lessen their presumably hurt feelings. The more time I spent with them, the more I projected myself onto them.

There's an attachment system that develops with humans and their offspring. It keeps parents around when children are screaming, doses the caregiver with oxytocin and other hormones when they're in the proximity of the object of their affections, and makes us perceive certain neotenous features—physical characteristics common to babies, like relatively large and widely spaced eyes—as cute.

"So how does it work in terms of pets?" Melson asked rhetorically. "Pets hijack this attachment system, so when they present neotenous features and signals of dependence, they trigger in us the same feeling as that adorable little puppy we want to hug. We are transferring what is in our DNA to make sure that we don't abandon our human infants to another species. . . . This very deep DNA attachment system . . . [is] like everything else in the human genome, highly malleable, highly flexible, and highly adaptive." My care for the chickens extends further than I meant it to, even without naming them. Their dependence on me is endearing, and while I cognitively knew it wouldn't be good to get too close to them, I couldn't help it.

CHAPTER 9

I think this is the misplaced source of my anger at Delaney: I have been hijacked by a DNA code that tells me to be fiercely protective over the chickens, even if they are destined for the dinner table. In some ways, her approach is probably healthier, and I recognize this on a deep level. Maybe I'm projecting some of the anger I feel toward myself onto her—*how could I be so gullible?* But also, since those emotions of attraction are very real, *how could I be so cold?* It's also possible that misery loves company; if I am going to slaughter these pets, I want everyone to be as devastated as I am. Either way, I want everyone to take this as seriously as possible, and I'm not amused.

In my experience, it's almost impossible to not take slaughtering chickens seriously, at least the way I choose to do it. There's something so visceral, so encompassing, and so intimate about the project that holding yourself at a remove doesn't generally work. During workshops I've led in the past, the people who trivialize the process, want to start early, make jokes, and project unequivocal confidence in their ability to kill chickens as if it's just *that easy* are the ones who have the hardest time sitting with the reality of the situation.

Slaughter isn't like chopping a carrot or cutting fabric for a quilt. It isn't casual. It doesn't map clearly onto unadulterated enthusiasm or an easy afternoon: there's too much life, too much of a nervous system, and too much bloody viscera that is too identical to what is inside all of us for that kind of emotional reduction.

When I was twenty-three and had only been slaughtering the animals I consumed for a few months, I bought seven roosters from a man on Craigslist. He had a backyard hobby flock and needed to do something with the males. I invited my friends Aaron and Seth to watch and help with the process. Aaron had been wanting to learn how to slaughter chickens for weeks. "Why didn't you call me to help you?" he asked after I killed and cleaned three ducks, then called to invite him over for dinner. I was sitting outside in my small yard, drinking tea and picking dried blood from under my fingernails, the phone pressed between my cheek and shoulder. My arms were slick with grease from the fat under the ducks' skin. It was a fair question: he spent more nights at my house than nights he didn't and kept a toothbrush in the bathroom. We were childhood friends who grew into inseparable adults. I considered him a brother and shared practically everything with him but wasn't sure about sharing slaughter. The dinners, yes. The killing, not quite.

"I'm not ready," I said. "I don't know how I'll feel with someone else there." I didn't want ego to come in; I wanted to focus on the animals. "But," I added, out of curiosity, "do you think you could do it?"

"I've already done it," he said. I knew that during college a few years prior Aaron autopsied a live chicken in a pre-med biology lab. They gave the chickens sedatives, performed their internal examinations, and then injected the animals with a second sedative in quantities large enough to induce an overdose. I didn't tell him that I suspected doing it in my backyard wouldn't be the same as doing it in the lab, that there wouldn't be gloves or masks or drugs on hand, or that I thought pushing a needle under the skin rather than drawing a knife across it would feel totally different, even if they produced the same result. I didn't say that participating in a science lab, wrapping a carcass in plastic, and throwing it in the garbage isn't the same as harvesting a living being for food and putting a body in the fridge.

During the rooster slaughter, months later, Aaron held a rooster for ten minutes with a knife on its neck before I made the cut for him. "At least I know I'm human," he said that evening, by way of explanation.

CHAPTER 9

During a large fifty-person workshop I taught one summer, we processed eight chickens and people worked in small, hands-on groups. I stipulated that everyone needed a buddy there with them during the moment of death, either to maneuver the blood bucket, grab a knife, or provide an extra set of hands. One man, Wiley, said he was ready almost immediately, right after sitting down with a chicken, and then added that he had no buddy and didn't need one. He grabbed the chicken roughly and didn't try to soothe her by rubbing her stomach before putting the knife near her neck. "Wait," I said, "like this." I found Wiley a buddy, slowed him down, and sat close enough to smell his breath. I pointed out the exact spot to make the cut. When the chicken started to bleed, although it was a good cut, although she remained still, although everything happened the way it's supposed to, Wiley started shaking. Tremors pulsed through his whole body, like his pulse had electrified. His legs and arms shook. Wiley's buddy held the bucket and took the knife, and I held both the chicken and Wiley. By the end he'd turned pale.

If blood and death render us all sensitive in the end, at least let me begin in that place of vulnerability rather than having those hard layers stripped off. Every time I've seen them go, they do not peel clean. I want to feel everything, to account for it, to hold that unbroken narrative intellectually and emotionally. As I fed the chickens the last of the grain, I thought about the strange phenomenon of being there almost every day of their lives and knowing that later that afternoon I would be there for the end. That I would be the end.

"I remember when you were just a little guy," I tell the rooster. A chicken crowds him to see what I'm talking about, and he pecks my hair again, which incites her to do the same. "I remember when I swabbed your butt and taught you to drink and mixed you dumpster beef puree and watched you dust bathe for your first time." I almost begin to cry, wishing they had names. I cannot tell which one is Jet anymore.

"Do you want some water or anything?" I ask Delaney, dropping my backpack off inside the door. "Come on in." The dogs circle her legs, eager to be petted by a stranger.

She says no, that she's good, but I barely hear her. "Are you hungry?" I ask. On the stove there are two pots of water and a kettle. I turn them all on to boil.

"No thanks, I just ate."

I grab a jar of nuts from the cupboard. In the last few minutes before Caitlin and Michael arrive, Delaney and I talk outside. She smokes a cigarette while I shuffle knives around the various stations I've set up and try to listen to what she's saying. The slaughter station is under the orange tree and includes a chair, a few scrap pieces of fabric to wrap the chickens in, an empty vegan yogurt container for collecting the blood, and an extremely sharp knife. The plucking station is our laundry line, adapted by tying six sections of rope with looped knots on the loose ends. The knots can tighten and loosen, allowing us to slide the chickens' feet through, secure them by the legs, and hang them while plucking. There are two paper bags waiting below on the grass for the feathers. The final station is evisceration. I have covered my small workbench (a slab of marble I pulled from a trash pile balanced on two secondhand sawhorses) and our outdoor patio table with black plastic, then adorned them with scissors, knives, freezer bags for the cleaned carcasses, rags, and labeled bags for various organs and body parts, demarcated "heads and feet," "intestines," and "heart, lungs, livers."

CHAPTER 9

Caitlin and Michael show up as the pots of water come close to boiling. The kettle screams. Everyone waits outside with the dogs while I bring the pots one by one to their various stations: the kettle to the evisceration table, for washing away the contents of a punctured crop or intestine, the two large pots, wrapped in towels to keep the temperature high, to the laundry line.

I take the group to the coop and we hand-feed the chickens dried mealworms. The chickens are outgoing and insistent. I want my friends to see the chickens like this, desirous and living, because this is how I see them.

"How are we going to do this?" someone asks.

"So, I'm going to show you how to bleed out a chicken. I'll sit down and hold her for a bit before, and then we'll take her over to the laundry line and get started on plucking. It depends on how many people want to kill chickens, and if you want to have people watching you or a little more privacy during the process. If you want to have privacy, then maybe everyone else can work on plucking while I help the person with the chicken. It depends."

"I don't know what I want to do," Caitlin says. Michael says he does not want to kill a chicken, and Delaney says that she wants to.

"Okay, cool. You also don't need to decide for sure until you watch me do it. But after we pluck all of them, then I'll demonstrate how to eviscerate them and get the edible organs out, and then I can help you guys work through cleaning your own. Sound good?"

It does.

I pick up the first chicken who comes up to me, a smaller hen who is always sweet. Even Amelia, despite her frustrations, often allows her to sit or dust bathe by her. It strikes me as ironic, and I wonder if the others feel this way, that I will kill her first for the simple fact that she trusts me and came running into my hands faster than the others.

I sit down with the hen and wrap a gray strip of cloth around her. "I think this helps her feel more secure," I say. "After she's lost most of her blood, she'll likely go limp, but after that their bodies sometimes shift and seize. The cloth helps keep them contained, because I like to hold them with my legs and abdomen so my hands are free."

I get into position, narrating as I go. The others stand and sit in a semicircle around me. "I like to have her body long," I say, stretching the chicken in my lap. Her legs elongate and hold there. Her back is against my left thigh, and her head hangs by my knees. I lean forward, my abdomen holding her between my legs. She is tilted head down, and I can feel the warmth of her body on my chest. She is much more like a basketball than Joan and Amelia, and her width makes her awkward to hold. I slowly begin to massage her. "Hold her head like this," I say, putting her skull against my palm, my thumb under her beak, my pointer finger over it. "You don't want to cover her nostrils," I tell them. "Let her be able to breathe." I move the knife to my side. I think of the other times I held her head to show her how to drink, to check her nose, to clean dried mashed potatoes from her beak, to admire the perfect circles of her ears and the way the feathers that clipped over the holes could pop open like snapdragons. The chicken relaxes. I feel her body ease into being held. I've held her like this before in preparation for this moment, for both our sakes.

"Chickens are like us," I say. "Their jugular is under their ear, below their jawbone. See this little flap?" I ask pointing to the short feathers covering a small hole, "That's her ear." I point to my own neck. "Where you can feel your own pulse, below your jaw, on the side—that's the same place where you want to cut on their neck." I

CHAPTER 9

watch Caitlin, Michael, and Delaney put their hands to their skin, to that soft animal pump below the surface. "You want to cut on the side of their neck, so you sever the jugular but leave the trachea intact. It'll allow her to breathe." I want to make this real. I want to align us with them. In slaughterhouses chickens are hung upside down, their heads run through electrified saltwater baths to stun them, and then a series of automatic blades cuts their throats. The blades sometimes miss. Deaths like that, ones that take place away from the public eye, are opaque and easily ignored. I do not want these deaths to be like that.

"How does everyone feel?" I ask, looking up at them. Delaney is sitting down, her hands over her mouth. Michael and Caitlin are standing, focused on me. "I'm going to do it." I tell them. There are nods.

I cut the side of her neck and blood flows. I catch it in the yogurt container and put the knife on the ground beside me. "She's limp," I say, "but I'm keeping a good grip on her." After a few beats, the majority of her blood is emptied by the action of her heart, and I feel her body begin to seize. Her organs shift and tense, her wings stiffen. "I can feel her body moving," I say. "I can feel the organs inside shifting and pulsing."

After the chicken goes limp, I remove the head and Michael helps me put it into a plastic bag.

"I can't do it," Delaney says. "I'm not going to kill a chicken today."

I carry the chicken by her legs to the plucking station and dip her into the hot water. "You want to make sure the water gets down to the skin," I say, "so you kind of slosh around." I dunk the chicken's body, using her feet to press her legs below the surface. The water turns tannish with the blood from around her neck and dust embedded in her feathers.

I slip one of the legs into the looped rope, then begin plucking, pulling the wet feathers with a downward motion from the knee toward her belly, wings, chest, neck. I collect the feathers into a paper bag, where they will dry for a week or so in the sun, so I can use them for pillows and craft projects later. Everyone stands back.

"It's so close. It looks like food," Michael says after I've plucked clean one breast, the legs, and one wing. Someone comments on the small holes in the skin. The chicken hanging from the laundry line looks like a chicken you might find in a refrigerator, or at least, it almost does. The feet are fat and thick with mud on the pads, and there are feathers still surrounding the neck and one wing. The skin around where the chicken's head used to be is flaccid and bright.

"Yeah," Delaney said, "back when I used to cook it looks like something I'd buy from a store." It only takes a few plucked feathers to make the animal look like meat, to transition from being to commodity, to find all of these other surprising, unfamiliar bodily features indicative of food instead of animal.

According to Wendell Berry, the link between animal and food, between land and animal, has been corrupted by modern systems of agricultural production and industrialized consumption. In "The Pleasures of Eating," Berry asserts that "the industrial eater is, in fact, one who does not know that eating is an agricultural act, who no longer knows or imagines the connections between eating and the land, and who is therefore necessarily passive and uncritical—in short, a victim." I watch them marvel at the simplicity of it: of turning a chicken into chicken, of reducing an animal, the live counterpart of a commonly encountered product, into something appetizing and edible, into something familiar. It seems so startling that something so basic to so many people's lives can be unfamiliar, that something so normalized as a product can appear so alien as a

CHAPTER 9

process. Delaney says that she had to sit down when I slit the chicken's neck because she thought she was going to faint, but now she looks at the body, up close.

I move the chair away from the blood on the grass in anticipation of spilling more. When I go to retrieve a second chicken, Amelia and Joan and the others watch me. I don't have to chase the hen: she knows me. Amelia and Joan are dust bathing and seem relaxed. They've clawed a series of holes in their run where they do something like wallowing, shuffling their feathers to get dirt close to their skin. Sometimes, they'll lie there in the dust for a while before getting up and shaking out, letting a large plume of dirt bloom behind them. Now, they're somewhere between actively bathing and lying there. It's always a little strange, how they don't change the way they look at me.

Although Caitlin has just seen it happen, I tell her how it'll feel to hold a chicken while it dies. "Her internal organs will shudder and move. Her wings will tense," I say. I know you can't ever really prepare someone for something like this, but I try. While Caitlin sits with the wrapped chicken on her lap, I crouch in front of her. The chicken moves, kicking her legs, and Caitlin exclaims. I reach out to help secure the bird while she settles.

Delaney and Michael watch from behind me. I get the knife, wipe it clean, and retrieve the yogurt container. When the hen relaxes, after I tap the knife on her neck to accustom her to the movement, and when Caitlin is ready, I make the cut below her jawbone. Blood flows into the container. After a moment, the chicken's body begins to shudder. The chicken kicks and shifts, her internal organs pumping nothing, her wings shifting. I hold her neck, slide my hand under Caitlin's arm, and elongate her legs. The hen's toes curl around my fingers and the soft yellow pads, fatty and warm, press into my knuckles.

"This is so beautiful and terrifying," Delaney whispers from behind us.

After a few moments, Caitlin says, "I can feel her vibrating." I do not tell her that I think the movement has stopped, and that I suspect it might be her own body. I watch Caitlin's temple and hands, put a single hand on the chicken's soft-feathered breast.

"I think this is done," I say, and we head to the laundry line. Caitlin, Michael, and Delaney help pluck this chicken, and soon there are multiple hands on her. The body is turned and picked at. The group finishes plucking quickly and efficiently.

Later, I start to eviscerate the colder chicken first while the second hangs, clean, from the laundry line. I position the chicken on its back, with its neck toward me, and pull at the tubes resting in their lattice of connective tissue below the loose skin of her neck.

"Right here," I say, "you can see the esophagus and the trachea." I cut the skin from the base of the neck to the end, pulling it back like a dissection. "The trachea goes to the lungs—it's the articulated, ribbed tube—and the esophagus goes to the crop. We want to get that out. I didn't feed them much this morning, so they probably won't have full crops. We'll be able to see the bugs we fed them." The trachea looks like the bendy part of a plastic straw wrapped in pale tissue; it is a pale, empty worm, almost like a more opaque version of an intestine. Where I cut it the organ bunches slightly, opening like a puckered mouth. I pull down to the base of the trachea, where it meets the empty bagpipe of the crop, then below where a tube connects it to the stomach. "Once you pull the crop free, use your fingers to get underneath it. Follow that tube and pull as hard as you

CHAPTER 9

can. You can either pull it free or cut it." I pull it out, long, and then cut as low as possible.

Next I turn the chicken around so her anus is facing me, and pinch the skin low on her abdomen. I make a small cut through the flesh there, telling the group to be careful not to go too deep and puncture the intestines below the surface. I push my fingers through the opening, into the abdominal cavity, feeling the smoothness of the still-warm organs, the slippery quality of the bluish looped intestines, and slide the knife from the inside out to cut the skin. "There's an oil gland here," I say, pointing to the small nub above the base of the chicken's tail. "You want to remove that as well." I scoop down with my knife to move the entire nose-like protrusion where the tail feathers once were, then work it around to complete the circle I started on the front of the chicken's body. Once I'm done cutting around the anus, everything starts to come free and I slide my entire hand in.

"Okay," Delaney says. The pale intestines fall free, steaming, as other organs—the heart, liver, gizzard—become visible while I hollow out the body. "I'm sitting down again." I know this means she's feeling faint. I keep going.

When I spoke with Dr. Melson, she told me she's a big believer in compartmentalization. "I think we all live with that and it's a very powerful part of the human brain. We're able to devote a lot of love and attention and nurture, and then at the same time get to a point to then kill the thing that we've nurtured. For some people it's difficult, but for many people they can very successfully compartmentalize." I scrape the lungs clean with the back of my nails, slice the gallbladder—that little green sack like a jewel of bile that will spoil the meat if punctured—from the rest of the liver, and show them how to bisect the gizzard, revealing grain and rocks, and how to tear free the yellowish plastic-like lining to produce a butterfly of smooth muscle. My emotions slide toward efficiency, toward pragmatism, toward protecting the body as a resource. I want to keep as much of the chicken as possible. Even if this doesn't enact compassion, at least it illustrates reverence. While it might be tempting to throw the organs away rather than clean them, I go through the process carefully.

I decide it's best to only slaughter two chickens with the group. Everyone is tired and has experienced, it seems, what they wanted to. It feels like we've reached a threshold. I assure them I can handle the other chickens by myself, and we sit around the patio table drinking water and talking about what happened. "You loved them like people," Delaney says, and I nod. She can tell I'm sad.

"It's always hard," I say, "but especially this. I was there almost every day of their lives." I feel like I'm missing something that I can't fully articulate yet. I'm also excited to eat the meat in the refrigerator.

"I've never seen something die on purpose," Caitlin states.

"It doesn't seem like a big thing buying them from the store, but when you see it happen it's something else. Almost everyone who eats meat has never seen an animal die."

Slaughterhouses are no kinder than backyards, and the Humane Slaughter Act fails to protect chickens. While he was serving on the Pew Commission, Bernard Rollin visited a slaughterhouse. "The animals were hung by their feet, heads down, on a conveyer belt," he later reported. "As they approached the killing machine, robotic paddles automatically fitted themselves to their heads, and an electric shock theoretically designed to stun them into insensibility was delivered." However, "the stunning failed approximately two-thirds of the time," so the chickens were taken through the next steps of the

CHAPTER 9

process—mechanical evisceration and a scalding water bath designed to make their feathers easier to remove—completely conscious. Every step of commercial slaughter is arguably horrific, from living on a CAFO, to being transported to the slaughterhouse, to being shackled upside down, to the stunning and scalding.

Delaney says she's still dizzy. Michael thanks me for including him. Delaney takes a sip of water and seconds his opinion. "It's so unlike anything else," she remarks.

Michael agrees. "It's really powerful." Caitlin nods.

Months prior, when I asked my friend Geran, who previously worked with a permaculture collective in Oregon, about the risks of backyard chickens as a practice of micro-resistance against dominant cultural practices around food and animal agriculture, he gave me a good warning. Geran knows a lot about the power of political movements: he's helped workers empower themselves with unions and spent time living out of his backpack to commit his life to activism. "Do you think this is an important form of resistance," I asked him, "or is this just way too passive?"

He took a breath before responding. "The problems that more personally focused, small-scale, permaculture-y movements fall into are in a lot of ways the problems that all small movements can fall into," he warned. "They become focused on just serving the individual and helping just one person. . . . The question always has to be, 'How do you give more than you're taking? How do you spread the knowledge and ideas and benefit from what you're doing?'" Workshops like this one give me a way to spread knowledge and a way to gift people experiences that help break down the modern disconnect between the production and consumption of animal products. If there is freedom in choice, that must start with awareness. Although I still have most of the chickens to process on my own, I consider the workshop a success. I am glad everyone came.

A few days later, I slaughter the rest of the Cornish crosses. In the interim I make fried chicken from one of the hens and almost cry. When I cut the legs free from the abdomen, watching the lock of a bone come clean from the joint that held it, I think about how the chickens ran and jumped, and then later, when they grew larger, how they clustered on the lip of the coop in the morning waiting for me to lower them to the ground one by one. I wonder, if someone were to fillet my hands and separate the edible from the inedible, what actions would be embedded in the muscle. The complexity of their bodies and simplicity of the action stun me; at base, I raised chickens in the yard and filled the refrigerator with meat without going to the grocery store.

Carol Adams, an author and animal activist, writes that butchering can help obscure in one's mind the idea that one is eating animals when one eats animals. By carving apart their bodies, humans can make a pile of meat no longer convey a sense of the animal it comes from. "Without animals," writes Adams, "there would be no meat eating, yet they are absent from the act of eating meat because they have been transformed into food." This allows people, she argues, "to forget about the animal as an independent entity; it also enables us to resist efforts to make the animal present." It's almost logically impossible to treat animals as individuals if you only encounter them as cutlets. Buying eggs or meat from a store removes the animal from the equation—when the animal is removed it's easy to forget where the eggs and meat once came from. Devaluing the role of animals in production and being nonchalant about the particulars of their life is a normal response to this system, but

CHAPTER 9

when animals are faceless abstractions, industries that maximize profits at animals' expense aren't pressured to calculate unquantifiable elements like their suffering.

But when you move from a live bird that blinks and scratches in the dirt and runs up to you eager for a snack to an oven-ready carcass, there is no obscuring this truth. Even the separated thighs in the fridge appear both as legs and food since the narrative that took those legs from body parts that belonged to day-old chicks to something I'd like to serve for dinner has remained unbroken and bounded to place.

Wendell Berry believes that the politics of food, like all politics, involves freedom. "We still (sometimes) remember that we cannot be free if our minds and voices are controlled by someone else," he writes in "The Pleasures of Eating." "But we have neglected to understand that we cannot be free if our food and its sources are controlled by someone else." This kind of control and awareness awakens a deep sense of autonomy and capacity. It's so easy to forget that the meat we buy from stores comes from animals when it arrives precut, devoid of blood, and wrapped in plastic. Cooking the meat, there are twin feelings of profound loss and profound abundance: the flesh is soft, wet, and rich. "One reason to eat responsibly," writes Berry, "is to live free." Their bodies, rendered to food, are a cosmos of clean plucked skin, connective tissue, and deep reservoirs of meat.

The morning of the final slaughter, I recreate the killing, plucking, and eviscerating stations, wear dirty clothes, and sharpen my knife again, soothed and mesmerized by the hiss of the blade against the triangle blocks I use. I test it on my fingernail, watching it scrape a small white layer of keratin free, before heading outside. I do not withhold food from the chickens this time, deciding that they should gorge themselves one last morning: I will be careful with the crops.

SLAUGHTERHOUSE IN THE BACKYARD

In the backyard it is bright and sunny, still early, the humidity relatively low and temperatures still down. I sit at the killing station and massage a chicken, a brave hen with wide breasts, and tap the knife along the side of her neck until she relaxes into the motion. I cut her jugular, bleed her out into the yogurt container, where her blood is warm and velvety. After hanging her from the laundry line, I go to retrieve another chicken and repeat the process. They assume I am bringing more food, so they run to me each time I return to the coop for another. I pick them up easily, and they relax into my grasp.

It might seem like a betrayal, but I do not think of it that way. Instead, I think that for their lifetime of knowing that whenever I approached them and held them it was a good thing—dried worms, throat scratches, dumpster scraps, fresh water, worms—this is the payoff. That their last moments aren't ones of terror. The blood comes and it comes fast, but until that moment they feel safe. I want to prolong that as long as possible.

After the chickens are all dead, I dunk each into scalding water and pluck their feathers free. They traverse the laundry line like small pale bats, hanging upside down from the ropes looped around their feet. I flash on the image that Isabelle Cnudde created for me months ago when I visited Clorofil: comparing the chickens hanging in their battery cages to small ghosts illuminated by the fluorescent lights of the henhouse. Their lives are very different from the chickens the rescuers left behind that night, although they ultimately both ended up dead.

I eviscerate the chickens efficiently: intestines into one bowl; livers, hearts, and lungs into another; gizzards into a third, where I bisect them one by one after all the chickens are cleaned. The last thing I do is cut into the grit-filled gizzards, for this step can dull a knife. Inside there are rocks, remnants of scratch and chicken feed, and small blades of crushed grass. I brush out the grainy debris and peel the yellowish plastic lining back from the dark red muscle.

I could say that I wouldn't like to be so attached, although I know I wouldn't rather have it any other way. I know these chickens are food, but I find myself spontaneously crying about their death during the next few weeks. I break down at two different coffee shops and in my shared office on campus. I cry at night when I'm trying to read.

I don't cry because I feel guilty about killing them; I cry because I miss them. I cry because it's sad. Because I've lost beings I care about. I cry because I wish I could still spend time with them, feed them, and watch them run with their strange bodies tipped forward, flapping their wings and waddling on their thick legs like tipsy women in heels who are late for engagements. I wouldn't be hiding spontaneous tears in crowded coffee shops if I didn't care for them, but I don't think I would appreciate eating them if they were just commodities. When I work with my body toward tangible results, my ability to remain aloof is eclipsed. Physical labor and emotional care invest me in projects, not just the other way around. I can't have an authentic connection to my food without becoming attached, and if feeling vulnerable and raw but not regretful is the cost of understanding this part of my experience, being able to consume meat, and raising birds that are taken care of, so be it. Let me be hurt and thankful, grateful, and conflicted. Animals are not plants. Slaughtering a chicken will never be like harvesting a carrot. I wouldn't want it to feel the same.

10

Waste Not, Want Not

THE OFFAL TRUTH

R ALPH WALDO EMERSON once wrote, "You have just dined, and however scrupulously the slaughterhouse is concealed in the graceful distance of miles, there is complicity." I can't argue with him, and I don't want to.

When you care for and slaughter animals yourself, this idea extends even further: you have just killed a being, and however easy it would be to avoid the mess of the offal, however straightforward it would be to only keep that pristine carcass, you have a responsibility to yourself and to the animal to consume as many calories from it as possible and thus extend the time before you slaughter another.

The nose-to-tail movement takes this responsibility and turns what some might consider waste into what it considers an opportunity. This kind of eating adopts the idea that we should consume the entire animal, not just the choice cuts, and that there's gustatory pleasure in doing so. It was not so long ago that our grandmothers or great-grandmothers were cooking liver and onions, but when farms became industrialized and chicken became something produced in factories instead of backyards, people no longer had consistent contact with viscera. However, according to Fergus Henderson, chef and author of *The Whole Beast: Nose to Tail Eating*, "If you are going to kill the animal, it seems only polite to use the whole thing." Henderson's book, which includes recipes that call for things like blood, spleen, heart, brain, and tongue, was featured on *Bon Appétit*'s 2017 list of "13 Healthy Cookbooks That Changed the Way We Eat." It was praised in the *New Yorker* as the "Ulysses of the whole Slow Food movement" and won the 2000 Andre Simon Award. The updated release of *The Whole Beast* included a foreword by Anthony Bourdain. It may be uncomfortable to imagine stuffing pig trotters or scooping marrow from bones—and perhaps less comfortable to imagine putting either of those items in one's mouth—but clearly there's something incredibly praiseworthy, and presumably delicious, about consuming the strange odds and ends left behind after the thigh and breast meat are spoken for.

What a Western diet considers "inedible" animal parts—like bones, blood, intestines, fat, feet, and lungs—composes a huge swath of any butchered animal. In cattle it accounts for just shy of 50 percent of the carcass, in pigs 44 percent, and in chickens 37 percent.

The industry often collects these parts, sanitizes them through heat treatments so they don't pose a significant public health risk, then disposes of them in municipal sewers and incineration plants. Quantities of these arguably unsavory harvests find their way into products manufactured for human and animal use. Fat rendered into tallow, for example, is often used in lipstick. Blood becomes fertilizer. Feathers, animal feed. Beaks and hair, pet food. Beef trimmings and connective tissue are heated, put into a centrifuge to remove the fat, treated with ammonia hydroxide to kill the virulent bacteria they harbor, and turned into "lean, finely textured beef." Consumers can, unhappily, find this in things like burgers and processed meatloaf. Bones with small amounts of meat on them are run through a sieve-like machine under high pressure to separate one from the other, ground into a batter, and called "mechanically separated meat"—which ends up in hot dogs and bologna (don't worry, the USDA limits how many bone fragments can be included in your meat paste, and since 2004, because of fears about consuming spinal tissue that carries mad cow disease, only chicken and pig carcasses still undergo this process).

Not everyone is willing to marinate their kitchen in the delicate aroma of simmering chicken intestines or add blood to their brownies, but the average American consumer—and their pets—eat strange animal body parts all the time. Defending our squeamishness ignores the fact that we come in contact with viscera and bones, regardless of the fact that their forms are rendered unrecognizable.

After the slaughter, I have an incredible amount of feathers, blood, feet, gizzards, livers, and intestines. I dry the feathers by keeping them in paper bags and setting them in the sun. I leave the bags open and insects consume any remaining flesh, and after a few days they do not smell and are completely clean and dry. I put the feathers into fresh bags and tuck them into a trunk alongside scrap fabric, where they remain until becoming pillow stuffing and cat toys. The rest of the edible matter has its own particular health benefits and possible applications, none of which involve ammonia hydroxide.

Although Western audiences generally find blood unpalatable, it's been used in culinary pursuits for a long time. Estimates suggest that the food industry only uses about 30 percent of all slaughterhouse blood, which is high in protein, iron, and particular amino acids. Even though it might seem unpalatable, blood can add texture,

CHAPTER 10

flavor, and color to dishes. For these reasons, blood lends itself particularly well to baking.

Blood and eggs have similar protein compositions and both contain albumin, a protein animals make in their livers, which gives them their coagulant properties. I've been cooking with blood since I began processing live chickens, and I've found that about a quarter cup of blood, maybe a little less, replaces a large egg. Blood, unlike eggs, can help fight anemia, one of the most common nutrient deficiencies, especially among women. It has a high bioavailability of heme iron and is metabolized by the body relatively easily, making the iron in blood a lot easier to absorb than the iron found in plants and leafy greens, like spinach.

Once blood is collected, it begins to clot almost immediately. While you can add salt to prevent this process, it will change the taste of the blood. A splash of vinegar works without altering the taste significantly. Forgoing either method, I let the blood clot when I slaughtered the Cornish crosses. When it's time to make cookies, I remove the container from the fridge and slide my knife into it. The blood, which has oxidized and coagulated in the refrigerator, is the consistency of slightly melted cheesecake or soft tofu. It is a rich, dark red, almost purple or chocolate in color, with a glistening, bright fire-engine-red layer on top where it was exposed to the air. The knife cuts it easily and I lift it out with a spoon. It holds its shape, jiggling, and I drop it into a bowl filled with sugar, vanilla extract, and oil. Although in the past I've broken up the clotted blood Jell-O by pulsing it on high in a blender, I've learned it's not necessary. I whip the blood and the other ingredients with a fork and they turn a delicate raspberry color. The grit of the sugar helps smooth the clotted blood.

After adding all the other usual chocolate chip cookie ingredients and dropping dollops of pink dough onto a cookie sheet, I bake them at 375 degrees for ten minutes. The cookies are rich brown, beautifully chewy, and slightly crispy around the edges. The chocolate and sweetness mask any kind of metallic tang. When Wendell Berry wrote that "eating is an agricultural act," he could have easily added that it can be a political protest. These blood cookies feel like activism because they argue that replacing eggs with blood is a fair swap. I would venture to say that, compared to eggs produced by factory farms, the blood in these cookies is a much more ethical product. Yes, this blood is the result of a knife, but it's also true that industrial eggs involve just as much death and arguably far more suffering.

The cookies don't last two days.

Some of what's left behind has more normal, culturally familiar applications. The bones, for example, go into a large pot of water and are boiled for hours on end to make bone broth. It's a new fad that's gained traction recently but very much an old idea (much like backyard chickens themselves). Bone broth was once part of prehistoric diets, and some early research on the diets of Indigenous populations in North America revealed that "bone grease," or marrow and gelatin extracted from bones and sinew after prolonged boiling, was an important source of calories. Bone broth is high in proteins, mineral salts, calcium, and collagen, which accounts for its comeback today. Proponents claim its anti-inflammatory properties can support joint and gut health.

I boil the bones until they can be crushed between my fingers.

Baked goods are one thing, but pâté is another. I don't like it, but I don't know this from experience. I've just never ventured a taste. At first it was because I thought it looked too much like canned cat

CHAPTER 10

food to actually be edible. Then it was because, right before I turned eighteen, I became a vegetarian. I've spent only a few months of my adulthood consuming the bodies of animals that I have not killed myself, and because of my prior bias against the pale paste, I generally reserved livers for stocks and long-simmering stews. I find the idea of mushed organs fascinating, but entirely unappetizing.

Not that it matters. It seems like the best and most direct to way eat livers, to turn them into a snack or meal unto themselves, is pâté. Chicken livers are high in protein, folate, iron, and vitamins A and B12. Livers contain selenium, an antioxidant that might help reduce the risk of diseases that cause mental decline, like Alzheimer's, and copper, which is essential for the formation of red blood cells and helps maintain healthy immune system functioning.

Since ancient Egyptian times, people have been force-feeding geese and eating their livers. Waterfowl have a biological predisposition toward storing extra fat in their livers, and this very physiological trick that helps them complete their seasonal migratory patterns also makes them susceptible to human intervention prior to making pâté. In order to get these dark organs to peak fat density and weight, geese and ducks are force-fed to their breaking point. By the time livers are harvested, they can be six to ten times their normal size. My Cornish crosses, voracious in their appetites and habits, were dieting by comparison.

Although pâté is often made with butter, some recipes call for schmaltz, or rendered chicken fat, and I'm a firm believer in trying to use what you have on hand. The chickens developed thick fat pads on their lower abdomens, on the stomach side of their vents. When I was done carving and cleaning them, I sliced my knife through the yellow, puck-like fat deposits, leaving grease on my fingers as I pulled them free. Now, I lay the golden hunks on the cutting board, chop them into smaller squares, and transfer them to a hot pan. As they sizzle and heat, the fat liquefies and collects, leaving small browned scraps of meat and tissue behind.

Once the fat has all but melted, I dice onions and let them cook until they are browned before adding a couple of cloves of chopped garlic. Although I am not particularly excited about the finished product, the onions, garlic, and chicken fat smell delicious.

The livers are next. They are a deep mahogany red, fresh-smelling, and slick without being slimy. Livers can be yellow, indicating a high fat content, or bright red. I lay them in a heap on the cutting board and remove the connective tissue with my knife. "Hey, dogs," I holler. There's a scampering of paws, and as they smell the raw meat and the cooking fat they sit immediately, tails wagging. Tashi enjoys the raw tissue, but Atlas spits his out on the floor.

I chop the liver into smaller strips and throw in into the hot pan. Almost immediately they begin to turn from a dark, hearty rouge to a softer brown. They snap and hiss, so I turn down the heat. As they cook, I open a bottle of wine and pour some into the pan to reduce. When the livers are done and the alcohol has burned off, I set the mixture aside, let it cool, and then transfer it into a blender. This step, to me, almost seems a little barbaric, like punishing an inmate by serving them nutraloaf. It's also called disciplinary or lockup loaf and is made by mixing meat, potatoes, vegetables, eggs, and whatever else is around, smashing it into a pan, and baking it (although some have tried to call this practice cruel and unusual, some states, like Arizona, still serve it while others, like New York, have banned it). That's what it feels like pouring the meaty, hearty, good-as-a-stew-base mixture into a blender: insulting someone's dignity by suggesting that paste constitutes a meal. I shrug, cap the blender, and press the container into place.

CHAPTER 10

After a few moments of high-speed whirring, I open the blender, scrape a small glob of pink-gray sludge from the side with the edge of a butter knife, and bring it to my mouth. Immediately, two things become clear to me. It needs more salt. And I have been missing out.

The pâté is earthy and rich, sweetened ever so slightly by the caramelized onions, made complex by the reduced wine and chicken fat. In the best of ways, it reminds me of the smell of wet concrete; it tastes like something dark and wet and warm. According to Julia Child, pâté develops its full flavor after a couple of days, but I don't plan on waiting that long. I put it in the refrigerator, set a timer, and begin to spoon it out with carrots for lunch once a few hours have passed.

It's not just blood and organs: there's also the paws to consider. While Americas eat an arguably colossal amount of chicken annually, the feet are often overlooked. When whole chickens are sold in stores, the feet are usually removed. Although some people add chicken feet to soups and stocks, which creates a broth so rich it becomes solid when refrigerated, chicken feet aren't necessarily a part of the chicken that many Americans think of consuming.

I have never tried chicken feet, nor have I added them to broth. Like many other Americans, I always viewed them as a waste product, even though my mom admitted to eating them as a kid. Perhaps this is from my socialization as an Americanized consumer and eater; perhaps it is due to the fact that chicken feet are such a rare item and so infrequently found on dinner tables that I was able to avoid them entirely.

The reality is that chicken feet are not a waste product but rather an arguable powerhouse of beneficial nutrients. Chicken feet are high in protein, calcium, and collagen. They contain glucosamine, which is good for joints. Research has shown that consuming collagen triggers the synthesis and reorganization of new collagen in our own bodies. Although collagen is most commonly thought of as something that's beneficial for skin health and complexion, the benefits go way beyond a nice-looking face. Collagen is good for our bone mass, density, and strength. It can support muscle health and slow cartilage degradation. Although chicken feet might seem undeniably icky, they are consumed all over the world. They add incredible flavor to soups and stocks, and considering the nutrients they contain, there isn't much of a good reason to throw them away. Since I can't logic my way out of eating the feet, trying them out is my only viable option.

First, I wash the feet in warm water and massage away the dirt and debris still on the foot pad. I scrub under the chicken's toenails even though I plan on removing them. Something about preparing to consume feet has me uncomfortable, and I scour each one as though it were infected.

After they are clean, I drop the feet into a bowl of scalding water to remove the skin. I strip it off, a yellow, almost plastic-like coating, in layers, starting at the ankles and moving down toward the footpads and toes, where it catches in the grooves. On the upper part of the leg the skin comes off almost like snakeskin, although some of the harder, thicker scales along the front of the leg require that I pop them off with my fingernail, one by one. Other sections, like along the back of the leg, pull free easily. After picking skin from the bottom of the foot, I peel the skin from the toes and leave it hanging around the nail. I chop each nail off with a strong knife.

Stripped of their skin, the yellow feet are white and slightly slick. They are still textured with the lizard-like scales and ridges of chicken feet, but stripped of their protective coating, they look vulnerable and bare, almost larval in color. I know it doesn't sound good, but I'm sure if I was socialized somewhere else or grew up with an understanding of chicken feet as food, the comparisons that come up might be more forgiving: the feet might look creamy or ivory-like.

Chicken feet are commonly eaten in Korea, the Philippines, and Vietnam, although in China they are particularly popular. According to poultry economist Paul Aho, chicken feet—or "phoenix talons," according to one of their much more regal names in Chinese—account for up to 75 percent of China's annual chicken imports. In China, chicken feet are often cooked two or more ways to soften their unpliable nature, and they are served either hot or cold in a variety of soups, main dishes, or even snacks to have with beer. They are deep-fried, braised, steamed, boiled, and simmered. The majority of chicken feet that were imported to China came from the United States until trade regulations and policies began interrupting the cross-country flow of feet.

In the past ten years, the United States and China have had a complicated relationship over chicken feet. The two countries historically closed their meat markets to each other to protect domestic markets and preserve international food security, limiting the risk of foodborne illnesses. In 2001 China joined the World Trade Organization, and the amount of chicken feet the United States exported to China grew rapidly and substantially. In 2009, when almost 80 percent of chicken feet imported to China came from the United States, the US was not accepting any chicken meat from China. After concerns about bird flu in 2004, the United States banned chicken meat from China, but in 2009 China reacted against this policy and filed a complaint with the WTO. "We have these jumbo, juicy paws the Chinese really love," asserted Paul Aho in a 2009 interview with the *New York Times*, "so I don't think they are going to cut us off," although China did put high tariffs on chicken feet from the United States, dropping imports radically.

Since then a series of issues, like continued tariffs on US chicken imports to China and the 2015 bird flu outbreak in the United States, have interrupted the flow of chicken feet from the United States to China, regardless of the fact that American chicken feet are, as far as I can tell, quite tasty and meaty.

I suppose I'll find out for myself. After each foot is skinned, I boil them in a small amount of water, vinegar, soy sauce, garlic, and ginger for over an hour. I stir the contents of the pot, adding more water as necessary while the broth reduces. Between bouts of sniffing and stirring, I make rice on the side. The liquid thickens as it boils down and the feet release gelatin into the mixture. The feet look tender and start to peel away from the toes and leg bones when I touch them with my spoon. They take on a dark brown color and an appetizing sheen.

After they soften, when the kitchen smells strongly of ginger and garlic and chicken soup, I let most of the moisture cook off until the feet are coated with a thick sauce. I pour the mixture, feet and all, over rice and toss a palmful of chopped green onions on top. It becomes clear that it's next to impossible to eat chicken feet with a fork, so I use my hands.

They are rich the way the best kind of chicken broth is and soft like nothing else. The residue on my hands and face is slick and sticky at the same time. The meat falls off the bone. Skin, tendons, and fat, all reduced to a gelatinous texture, practically dissolve in my mouth. If everything was stripped clean, it would be possible to spread the collagen-rich almost-paste over toast with a butter knife. I bite off entire toes with a soft pop, then massage them around my mouth with my tongue until the meat falls free and I can spit out the small columns of bones. It feels oddly reminiscent of extracting watermelon seeds from the soft, wet fruit. I am immediately sold on chicken feet and make them again within the week. Although I have yet to try intestines—the final step, it seems, in the process of consuming an entire chicken—I feel confident, considering how well things have gone so far.

Some people think that there is no such thing as an ethical omnivore, but I do not necessarily find the term oxymoronic. There are ways to eat meat that are not so tied to industry as other ways. There are ways to sidestep (at least partially) capitalist forces and do less evil. They exist.

But if there is to be some kind of omnivore's ethics—or any eater's ethics, for that matter—I would suggest that it involve a level of consistency and vigilance. It's clear, at least according to Geran, that "obviously the animal industrial complex is a fucking cosmic sin," but where does that leave us and our choices?

Perhaps eating isn't always activism, but if we treat it that way, we open ourselves to the possibility of acting in accord with our beliefs at least three times daily—or six to eight times daily, for those of us prone to snacking. Or maybe when we allow eating to feel like activism there are even more opportunities to reify our beliefs with action: maybe then picking bugs off cucumbers, composting chicken bedding, turning over rocks so the chickens can find worms underneath, maybe all of these actions become something other and more than what they were. They become the commitment to do good and the hope that we can.

If you've taken a philosophy class, you've probably heard of Immanuel Kant's categorical imperative. It's more of a moral yardstick than a set of rules like the Ten Commandments are. "Act only according to that maxim," Kant suggested, "which you can at the same time will that it should become a universal law." Staring at the warm intestines, I sigh. I know that if I want something, I need to work for it. I want there to be less waste in the world. I want fewer animal products to be thrown away since, as Fergus Henderson suggests, that is the only polite thing to do. And I want to do less evil.

I resign myself to the intestines. Although I've never seen them on a menu or dinner table, they are not inedible. In the Philippines, *isaw* is a common street food dish. It's made of pig or chicken intestines that have been cleaned, turned inside out, boiled until tender, skewered, grilled, and flavored with chili and vinegar. They're often paired with a beer or served over rice. *Adobong isaw*, or chicken intestines, are stir-fried with garlic, bay leaves, peppercorns, ginger, chili, lemongrass, vinegar, and soy sauce. It seems vulgar to come this far into the world of unconventional eating, at least by Western standards, trying things like crickets and blood cookies, and not make use of the intestines.

With a certain degree of trepidation, I bring a bowl of intestines to the sink, cut them into six-inch-long strips, and begin squeezing the feces and partially digested matter through the open ends. What comes out is a combination of mashed food, most recently peaches from a dumpster, grass, bugs, and a small amount of organic, conventional feed. The grassier pieces are at the beak end of the digestive track, while at the vent end the matter becomes an indiscriminate green the consistency of slimy, putrid mayonnaise.

I cut them open lengthwise with a small knife, rinse away any remaining fragments, and then realize there's a strange paste stuck to the inner side of the white, fleshy strips. It's gelatinous, smooth, yellowish, and musky. With the side of my knife I scrape it free in a homogeneous mash. I rinse the intestines a second time and then a third. Finally, after soaking the bisected intestines for a couple of hours in vinegar and saltwater, hoping this leaches out any lingering fecal flavor, I drain the intestines from their cleansing soup and heat a small amount of oil in a pan and sauté onions, garlic, and ginger.

I hold my nose close and inhale deeply. They don't have any odor besides the slight snap of vinegar, and I am grateful that is the only scent I can detect. I pour them into the pan (the onions have browned and the ginger and garlic become wonderfully aromatic), cover them with water, and add another splash of vinegar for good measure. As the water heats, the intestines turn from pink to white, translucent to opaque, and curl into themselves like pig tails. They look like some form of sea creature, and I recall a distant memory of being five or six and refusing to kiss my mom because she'd been eating tripe, cow's stomach lining simmered in spices and red sauce. I add dried chili flakes and a drizzle of soy sauce.

My five-year-old self would not have entertained the idea of intestines, but my older self finds the smells of garlic, ginger, chili, vinegar, and something strangely, slightly sweet as unnervingly attractive.

Over the next forty minutes, I cook the intestines until they become tender. I keep adding water and stirring them, scraping the browned sides of the pan into the middle. Every so often, I try them. Although they start out rubbery, by the time I pour them into a bowl they're palatable, with a consistency that's a little more forgiving than calamari. The deep amber color is strangely appetizing.

With all the seasonings, it's hard to pick out a distinct intestine flavor, but I consider this a perk. The sauce that formed in the pan is thick, spicy, and salty. It tastes like something I'd eat to fend off an oncoming cold—all that ginger and chili. The slight resistance of the intestines contrasts impeccably with the soft onions, which practically melt upon contact with my tongue. Although the process itself was a bit disconcerting with the squeezing, rinsing, and scraping, I can't pretend that I don't enjoy my meal.

It seems to me that omnivores should be morally obliged to at least consider consuming animal products if for no other reason than it can enhance their culinary experiences. That's a selfish, small reason, although I'm not sure the lofty one of reducing waste and drawing more meals from a single creature would necessarily convince the masses. During one of our phone conversations, Nina, the caterer I picture as a rebel foodie, explained why she thinks purposeful, considerate purchasing practices are important. "I'm a vegetarian," she told me. "The more vegetarian products are purchased, the more people realize there's a market. When you create the demand, there's more of it. Similarly, going out of your way, if you do eat meat, to buy free-range, humanely raised stuff is a good way to do it." Just as buying meat-free burgers can increase the demand for and availability of vegetarian products, being willing to try dishes with offal and blood, whether you're buying bone broth in stores or asking your butcher if they have available organ meat, increases the demand. Utilizing more of every animal means increasing the number of nutrients, the amount of protein, and the caloric sums that can be gleaned from it. Seventy percent of the blood that flows through slaughterhouses ends up in landfills. I think about these things as I organize my freezer, the colors of my jarred collections as eclectic as any candy store: mahogany-red blood, nearly purple gizzards, rich plum livers, pale pink intestines that are only a shade lighter than naked ladies, a type of lily that smells like bubblegum.

11

After Harvest

TO GET TO THE OTHER SIDE

About a month after I slaughter the Cornish crosses, I move to a semi-rural neighborhood and throw a housewarming party, and about twenty of my friends show up. In the back room there are two baby ducks in a brooder, six fish in the aquaponic system that grows thyme and Thai basil, and a closet that has three brooders housing a collective total of eighteen rainbow ranger chicks—meat birds, once again—living under red heat lamps.

In the yard I have a garden with cucumbers, tomatoes, peppers, and strawberries. By the front porch, laden with pots of aloe, mint, and flowers, I build raised beds and fill them with pole beans, oregano, cilantro, carrots, and sage. Behind the house, Joan and Amelia live in an expansive chicken run with grass, trees, and an abundance of bugs.

Every morning I clean all four brooders, change everyone's food and water, put my coffee grounds into a worm bin I made to compost kitchen scraps, let Amelia and Joan out of the coop, feed and water them, take the eggs inside, and carry a five-gallon jug back and forth three times to the way-back garden. I water the plants in the front, then sweep up the pine shavings I've inevitably scattered throughout the house.

I fall into the regular rhythm of early morning chores, trying to finish everything before the light creeps over the mimosa and walnut trees that border my yard. It is again summer and viciously sweltering in southern Louisiana. I wake up early and go to bed early. Every morning I start my day with a reminder of the resources I use, with compassion toward my animals and the burning feeling that this is what it feels like to put my own hands to good use.

There are so many ways to find belonging and so many ways to lose it. Everyone needs to eat, and the status quo rests on Styrofoam-and-plastic-packaged meat and processed foods, on high-waste lifestyles, on buying instead of laboring, on exchanging time for money for goods rather than pulling necessary sustenance from the soil. If I am to belong—if any of us are to belong—the channels through which we nourish ourselves must be intelligible to our human hearts and animal bodies.

If the world doesn't change, we might be headed for a global catastrophe. For every hundred calories of grain a chicken ingests in factory farms, they only produce twelve calories worth of edible meat. It is currently almost common knowledge that the global livestock industry creates more greenhouse gas emissions than all transportation—that's planes, trains, ships, and automobiles—combined. In an article by Damian Carrington, Rob Bailey, lead author of a 2014 study that analyzed the effect meat and dairy consumption has on climate change, said that "preventing catastrophic warming is dependent on tackling meat and dairy consumption, but the world is doing very little. . . . A lot is being done on deforestation and transport, but there is a huge gap on the livestock sector. There is a deep reluctance to engage because of the received wisdom that it is not the place of governments or civil society to intrude into people's lives and tell them what to eat." The problem is that if no one starts telling us what to eat, there might not be much left to consume. A 2014 study published in the peer-reviewed journal *Nature Climate Change* showed that current trends in yield will not sufficiently meet the projected global food demand in 2050, and we won't be able to feed everyone without considerably expanding agriculture. The problem? Agriculture already covers much of planet Earth, with animal agriculture dominating about 40 percent of the world's arable land. Animal agriculture is one of the main contributors to not just climate change but also pollution and loss of biodiversity.

There are alternatives. Part of the problem with industrial agriculture is the scale upon which we do it and the way we do it. Backyard chicken owners enact a different set of ideals that run counter to the narrative that we should consume what is produced for us. By reminding people about their connection to the land and encouraging them to bond with animals that can feed them, the backyard chicken movement is a widespread feat of micro-resistance against systems that habituate people to staying hungry and disconnected.

Often our human desires for food, sex, and recognition are commodified, exploited, marketed to, and pathologized. Society's recommendations about how to navigate these needs with things like fast food, social media likes, new technology, and more stuff partially fulfill yet fail to genuinely satisfy. These half solutions both distract and malnourish us just enough to keep us struggling for more, thinking these replacements for food and connection could one day be enough. This makes the plight of others and the environment seem a little more out of the range of our perception, a little more unfocused. In the words of Adlai E. Stevenson, "A hungry man is not a free man," and we are becoming a nation of hungry consumers who cannot tell right from wrong because those two options are often irreversibly interwoven. By coming back to what is real and what is useful, we can find what is ours. A chicken is not just a chicken.

The onslaught of want, which can be quelled by cheap goods but will never be satisfied by them, anesthetizes us to the feelings and values of others and makes us oblivious to our place in the world. Fulfilling large desires, like the need to belong and be nourished, with easy and therefore insufficient offerings, like posing in trendy clothes or eating food that destroys our bodies and the earth, teaches us that we are small.

It's so easy to participate in systems that are morally questionable because we have been taught that we are not individually significant. This again reminds me of the concept of naïve cynicism—a mindset that eschews complexity for simplicity, often inaction for action, all the while maintaining its mature superiority. People can become petty enough to think that what is convenient one moment will make them happy later and delusional enough to think that because they are just one person their actions won't change anything. People are not consumers first and foremost: we are producers. Humans are vibrant and intelligent and creative, and we desire engagement and contact with the real. When we lose touch with that bodily capacity to create, to be self-sufficient, and to give unto ourselves, we forget who we are. We become dislodged from our place in the world. I sometimes think of Kathleen of Moonwater Farm and what she said: "People are starting to understand how food is medicine and that it is a source of agency when we can grow our own food."

Anything that reminds us we have agency is worthwhile. This, if nothing else, is one of the simplest lessons of the backyard chicken movement: Individuals can raise livestock and produce eggs just steps from their kitchen. They can care for an animal that feeds them. Anytime society uncovers a widespread reminder of agency, it means something. Like a divergent fault line that breaks the earth open, a low-pressure area is created in the wake, asking for something to fill the space.

Perhaps this is why homesteading movements seem so prevalent today. In addition to more farmers' markets than ever before, there's growing access to and awareness of organic and non-GMO produce. People are rewilding and relocalizing. If the backyard chicken movement did not have so many corollary movements that support the same ideals of food security, environmentalism, self-sufficiency, and labor, perhaps one could argue it isn't consequential.

The backyard chicken movement is not an anomaly. It doesn't seem to be just a random quirk of some cultural zeitgeist catalyzed by anxieties about urbanization and fantasies about homesteading. The backyard chicken movement, in part, came from a building cultural pressure that mythologized pastoral ways of life and saw "soft" forms of labor, like cleaning a coop, changing water dishes, and collecting eggs, become suddenly acceptable, even desirable. Yes, pressure can build and dissipate, and the backyard chicken movement might not result in any lasting or meaningful change. It could just be a distraction, or a kind of emergency release valve that takes the pressure off our anxieties about lack of connection with nature and food without substantially shifting the way people eat. It's possible that when hens live in America's backyards, society will be less vigilant about the ways agribusiness and urbanization have changed how *we* live.

It's also possible that backyard chickens might act a little like green capitalism. Both can act as the creation of an alternative that makes it easier for the conventional system to perpetuate itself because it seems like it's possible to step outside of it—or as a choice that is so steeped in privilege it might create blind spots around the reality of systems of economy and food. There is an undeniably middle-class element of green capitalism, or of making particular products in a capitalist society seem a way out of our current problems of land pollution, worker exploitation, and animal welfare abuses. There are a lot of ways certain products we buy are marketed as more environmentally sustainable than the alternatives. While it is definitely good when companies adopt fair practices of worker treatment and create things that use fewer resources, have more sustainable systems of production, and result in less disruption of ecosystems or fewer instances of animal rights abuses, these systems do support our general status quo of production and consumption. They benefit those who can afford such products and risk creating a moral hierarchy surrounding purchasing habits that are grounded in social inequality rather than ethics.

According to Ariel Salleh, an ecofeminist who is concerned with the connections between nature and capitalism, "A light-green middle class can coexist quite comfortably with capitalist despoliation of the world, because it can afford to eat organically grown food and buy houses in unpolluted places." If people live with a level of privilege and disposable income, it becomes possible to purchase products that are marginally better than conventional ones. It can be easy to forget that the alternatives are only alternatives for some people in some situations. It's also possible that an alternative—like backyard chickens—can end up feeding back into the system that it seems to want to escape.

Backyard chickens are still a little counterculture, but they are also not so far from the mainstream that they can't be absorbed into it. People might want to escape commodification of goods, like eggs, by having a direct and meaningful relationship with the animals who lay their breakfast. But then again, backyard chickens can also be commodified through things like trendy Instagram hashtags, chicken tutus, and fancy chicken accessories, essentially turning them into a more modern equivalent of the television that is meant to signify you're successfully keeping up with the Joneses.

So maybe the pressure that catalyzed the backyard chicken movement will dissipate, but here's what's also true of pressure: if it's great enough, it always changes the things it exerts force over. Even if pacification and escapism—twin tools of the systems that led us to a place where most people who eat eggs have never held a chicken—threaten the backyard chicken movement in the future, the pressure that made urban chickens a widespread possibility and something that people want to incorporate into their daily lives changed our society.

In a world in which society was not so unconcerned with the animal welfare, environmental, and human rights abuses of industrial animal agriculture, the amount of suffering the poultry industry causes wouldn't be conceivable. By contributing blindly, trusting corporations, and allowing industries to perpetuate the myth that we fundamentally are consumers rather than producers, things that were previously unthinkable have become a reality. But just like unconcern can lead to new types of suffering that can lead to atrocities, waking up, looking around, and realizing that somehow our values have changed or don't align with dominant cultural practices can lead to feelings of pressure. This cognitive dissonance between what we feel or want and what we do or how society tells us to live can lead to solutions. The silver lining of social problems is that it makes solutions to those problems, and not just the issue itself, thinkable. The carelessness that makes atrocities possible also, eventually, makes the way out possible.

Backyard chickens are a way out. They are a small act of resistance, a slight but meaningful shift, a mainstay against the separation of production and consumption in America's food systems. The pressure that catalyzed the movement, like a force that transforms coal to diamond, changes people.

Micro-habits can also change people. Working for the things I want helps me feel more at home in my body. Knowing what goes into my food gives me the autonomy to make choices that I agree with more frequently. Trying to do better and knowing there will not be a single, simple way to accomplish this helps me resist the pull of naïve cynicism and reminds me that the perfect is the enemy of the good. Treating food as activism opens up new possibilities for protest and gives me a chance, three times a day, to vote for what I believe in. Refusing to expect an easy way out reminds me that this is a process that won't get simpler because that is not the nature of food and politics, nor does it need to be. Struggling to find more authentic ways of being doesn't have to be unpleasant. Micro-habits can fundamentally change the way we interact with the world and ourselves.

I wonder where I will go from here, where we all will all go from here. There are things you can't unsee and lessons you can't unlearn. As Matt from Carolina Coops told me, "One day my chicken girl said to me, 'These chickens will be worth more than diamond earrings one day.' She was right." Even if society doesn't collapse, even if the environmental apocalypse doesn't come, laboring in a way that reminds you of who you are and how to make decisions for yourself is valuable. Even if they are not the only food available, backyard chickens can be worth more than diamonds in a world mesmerized by cheap sparkles because backyard chickens can be a reminder of what's

CHAPTER 11

real. They illustrate the possibility for a relocalization of food production and can tie us directly to the processes of our sustenance. The small lesson that eggs come from chickens can expand into a greater realization that food is something alive and not just a supermarket commodity.

When I spoke to Isabelle at her micro-sanctuary, we talked about the future of food. "I feel like there is a lot of growing awareness about what the egg industry is actually like and what chickens go through. I think people are talking about it more," she said. "What," I asked, "do you think the future of the industry might be? Do you think people will change their habits?"

Isabelle blurted out, "That's what I want," before pausing. She continued, "Whether that's going to happen, that's the question. I'm not sure about people. I feel like there's still a lot of awareness to happen." She says the fact that people are still surprised about things like egg-laying hens being killed at eighteen months, that broilers have fundamentally different body structures than layers, and that hatcheries commonly kill baby chicks means that even chicken people aren't always wise to the industry and its practices. "I don't know what the future is going to be," she says, "but I hope that there will be a lot of new food technology replacing eggs. Because there already are many, many ways to replace eggs in whatever you cook and bake."

Even though there are replacements, that doesn't mean everyone wants to use them. The problem is that most people in the United States are accustomed to the idea of buying eggs in stores and not thinking about the farms that those cardboard cartons came from.

"I think it is so hard to change people's eating habits," Isabelle said. "They are so entrenched in the culture. I am really hoping that it is going to be replaced without people knowing it, that all the stuff using eggs is just not going to have eggs anymore, the products are going to change and the people are going to consume them without really knowing."

Her husband, Peter, sitting beside us during this conversation, chimed in with what he thought. "If costs change," Peter said. "If some of these externalized costs get put back on the producers. If, therefore, prices increase because laws get passed, and laws get passed for welfare reasons." Lots of eggs don't get eaten as eggs; they get eaten as baked goods and protein bars and packaged food.

If we don't change, we might be heading for catastrophic failure. "Antibiotic resistance," Peter said. "We are in real trouble. And it's all directly related to keeping so many animals so close together, which forces you to overuse antibiotics. I mean, there's no other way to do it. Given all these trends," he concluded, "I think that it's inevitable that at some point we'll go to a maybe not exclusively plant-based diet, although that is what I would want, but to mostly plant-based foods."

Isabelle said she thought that it's hard to get people to change their habits, but Peter disagreed. "People change their habits all the time. The food we eat today is not the food we ate fifty years ago."

Hilary Near, a commercial zero-waste assistant for the city and county of San Francisco, also believes habits can change. "If we can get to the younger generation sooner, instead of systems that are easy for them to participate in, then we have a chance." This means planting environmentalist seeds young and teaching kids about reducing their waste and living in ways that put less pressure on the environment. "If adults have been doing something one way," Hilary said, "it's harder to change that behavior. I think the power of governments and service providers and entrepreneurs to set up systems that make it easy for people to use reusables is huge." In the future, since our current practices are so unsustainable and people are

CHAPTER 11

starting to wake up to the idea that things need to change, there will hopefully be more pressure on governments and companies to behave in more rational ways versus more capitalist ones. This can have serious ramifications for cultural expectations.

"Hopefully, the next generation takes it for granted that they don't even leave the house without their coffee cup and reusable straw, if they need one," Hilary says. "Soon, if you see a plastic straw, you'll be aghast, like 'Oh, what is that disgusting thing?'"

The backyard chicken movement promises that perhaps the food of tomorrow will be more akin to the food of the early twentieth century. It challenges the idea that people cannot produce their own food by making chickens into pets and companions. I imagine that I will always keep hens in some form or another. I suspect I will remain a dumpster diver.

When I asked Nicole and Chicken Mike why they thought backyard chickens were becoming so popular, he said, "It's love. Love is why it's spreading."

"It might start with health," Nicole said, "but then people realize that chickens are fun." She compared it to the peaches she grew from her tree that year: the branches were laden with fruit, and when some guys next door working on a roof yelled, "Can we have some peaches?" Nicole called back, "Yeah, how many do you want?" They were surprised and partially kidding, but maybe if they hadn't noticed how much work had gone into the peaches, how ripe they were, how cared for they seemed, they wouldn't have ventured the joke—which always has at least a little grain of truth in it. Nicole attributed her peach harvest to the chicken manure she spread at the base of the tree.

"That kind of positivity spreads," Mike said. "Maybe the guy will grow peaches, maybe he'll share them with his neighbors, but either way he saw how good those peaches were and probably told his friends about it. Maybe it'll inspire him to grow peaches. Maybe peaches will be his thing." Like homegrown peaches, chickens take love and give it back. People feel proud of homegrown produce and eggs. It's only natural they want to share, and when people taste those superior foods, it's only natural they want to produce some themselves.

The love in the backyard chicken movement is palatable. Dave loves Sammi. Isabelle loves Clorofil. Jenny loves her pets, and Kristen loves her brood of silkies. Matt loves making his Carolina Coops. You have to love what you do to keep doing it without destroying yourself, and although backyard chickens are often considerably bougie—look no further than the catchy Instagram hashtags and chicken accessories—the beating heart of the movement runs off love and a prayer. It's enacted over and over through labor, through the belief that things can and will be better, micro-resistance is a fair way to fight global corporations, and people have the power to produce their own food.

At the end of the night, I send people home with party favors of roast chicken, fried chicken, and pork ribs that I rescued from a dumpster. Passing off free meat, I know that without this step all of these animals would have lived, consumed resources, and died needlessly. "Waste is a verb," Max from Urban Ore told me, and I understand that *valuable* and *useful* are verbs too.

There is no telling what will happen next. Some suspect bees will become the new urban chickens and that honey will replace eggs as the ultra-hip thing to manufacture in America's backyards. Even if

CHAPTER 11

the movement doesn't expand or translate to other forms of urban livestock—although goat yoga does seem to have something going for it—there's lasting worth in this particular moment. No matter how these desires continue to shift in the future, it seems like society has moved to a point of recognition that it's not going to back down from.

By the time the partygoers are gone, my casket freezer—so full at the beginning of the evening that I couldn't close it and had to move ice to a separate cooler—is almost empty except for some as yet unused remnants of the Cornish crosses (meat, blood, feet) that I know will eventually be put to good use. It seems to me a small miracle that so many people would be so willing to take the salvaged chicken meat home and so understanding of the need to.

About a month later my parents visit as summer is drawing to a close and the chickens from the closet have been moved into a mobile pen in the backyard and the ducks into the coop with Joan and Amelia.

My parents are amazed by the size of the mobile pen, although they've seen pictures. I acquired the wood from dumpster diving at construction sites, stopping on the side of the road when I saw an attractive plank or plywood board, and from a friend whose company had unused spare boards. It's an unruly thing, partially because my ambition for scale wasn't tempered by the realization that dragging it across the grass would be nearly impossible, even with the small wheels I added as an afterthought. It's five and a half feet wide by twenty feet long, with a sleeping loft that's five and a half by six feet. In the loft, there's an elevated perch and a three-foot-high ceiling to allow ventilation on hot summer nights. The chickens learn to walk toward the new grass as I muscle the structure forward every couple of days. The rainbow rangers peck at the bugs, who jump and flee as the crowd of small dinosaurs charge forward as fast as they can. The rangers are so tame that I often open the door of the chicken run and let them wander around the yard; they come when they are called, and if that fails, they come when I wave one of their food dishes above my head. They are willing to eat almost anything that I feed them, and because of their flexible palate I'm able to feed them primarily off dumpster food.

During the first night of my parents' visit, I pull one of the last Cornish crosses out of the freezer to prepare for dinner. After letting the hen defrost, I fill her internal cavity with carrots, onions, potatoes, and rosemary and slide sliced garlic cloves mixed with olive oil, basil, thyme, lavender, and fennel under the skin. It roasts on low for five hours. I make biscuits from scratch and salad dressed with homemade kombucha vinaigrette.

We sit down at the table. Candles are lit, water is poured, hands are held. We thank each other for everything we have done. They thank me for the raising, slaughtering, plucking, and cooking, and I thank them for the visit, the support, the temporary coop over the winter holidays, the understanding, the endless love.

They have never tried homegrown chicken before, not like this. Mom is amazed at how juicy and flavorful it is. It is a chicken that actually tastes like chicken. It lived its life running, eating bugs, consuming grass, and pecking in the dirt. What complex ramifications for the simple act of letting a chicken exist as a chicken. Dad helps himself to seconds. There's something incredible about the ease of this meal, the way it reifies what I believe and how I want to live. As I watch my parents enjoy their dinner, I feel something deep and warm. Pride. I am proud of what I accomplished.

Belonging is a verb too. There is no way to be a part of something without earning it. Autonomy in consumer culture is impossible

CHAPTER 11

without labor because it is so easy to have all of our decisions made for us. So often others decide what we should eat, how much it should cost, what ingredients go into it, and what environmental impacts are caused by its production.

After dinner, I walk to the mobile chicken pen in the backyard. The rainbow rangers have crawled up the ladder into the second level and are arranged on the perch.

People are radicalized incrementally. At first it's just a few hens, and then it's eight broilers, and then eighteen in a massive mobile pen designed to provide them with fresh grass and insects. And that's all it takes. There's some kind of rationality in local food and acts of labor that makes going back seem unfathomable.

After opening the long side door, I lift the ladder and lock it in place to protect the chickens from predators. I tell them goodnight and walk back toward the coop where Joan, Amelia, and the ducks are already waiting inside.

The yard is buzzing with summertime insects, and on the way back, I can make out the planter boxes holding the skeleton frames of tomato cages and the cucumber trellis. Inside the house, I see my parents walking past the windows, laughing. On the top floor of the coop, the ducks have situated themselves in the nest boxes while Joan and Amelia are side by side on the perch.

It has been almost exactly a year since I built the coop, ordered chicks online, and decided to become a backyard chicken person. Over the course of that year I met chickens rescued from egg farms, productive chickens, backyard chickens, chickens who live in fancy coops, chickens who live inside, chickens who ride in private jets, a therapy chicken, and a swimming chicken. I came to occupy two ends of the backyard chicken owner spectrum—someone who travels cross-country and camps with her pet chickens but also someone who eats her meat birds—and I met the CEOs, activists, urbanites, fashion designers, parents, and homesteaders who occupy all other points on that continuum of chicken ownership.

This is not just the story of backyard chickens. This a story of a fringe social phenomenon becoming arguably mainstream as it reimagined bygone ideals and challenged modern ideas about food, eating, and animals. It's about how hearts can open to even the most unlikely of creatures and how so many of us house a self that feels trapped under fluorescent lights, craves something beyond a microwavable meal, recognizes the beauty in labor, feels at home in the elements, and senses the implicit reason in sweat.

The story of backyard chickens cannot be told without telling the stories of the people who work toward reclaiming autonomy and agency within their consumer identities, who choose to invest in what's right and worthwhile rather than what's easy, who make their world natural and wild and sustainable so they can find themselves within it.

The only way to belong is to work for it. I put my hand between Joan and Amelia to make more room between them on the perch—it's a hot night and they sometimes crowd each other. I know they will spend the evening pooping in the nest boxes, and I know I'll just need to clean them tomorrow morning, but I don't mind. I cup a hand around Joan and then Amelia before closing the coop doors, sliding the latches in place, and heading back inside.

BIBLIOGRAPHY

GENERAL SOURCES

Bartling, Hugh. "A Chicken Ain't Nothin' but a Bird: Local Food Production and the Politics of Land-Use Change." *International Journal of Justice and Sustainability* 19, no. 1 (2012): 23–34.

Earth Policy Institute. "U.S. Meat Consumption per Person, 1909–2012." March 7, 2012. Excel data available at www.earth-policy.org/data_high lights/2012/highlights25.

Elkhoraibi, C., R. A. Blatchford, M. E. Pitesky, and J. A Mench. "Backyard Chickens in the United States: A Survey of Flock Owners." *Poultry Science* 93, no. 11 (November 2014): 2920–31.

National Chicken Council. "U.S. Chicken Industry History." 2019. https://www.nationalchickencouncil.org/about-the-industry/history.

Rollin, Bernard. "Raising Consciousness about Chicken Consciousness." *Animal Sentience* 17, no. 2 (2017). https://animalstudiesrepository.org/animsent/vol2/iss17/2.

CHAPTER 1: FLOCK FRENZY

Berry, Wendell. "The Pleasures of Eating." In *What Are People For? Essays*. San Francisco: North Point Press, 1990.

Bouvier, Jamie. "Illegal Fowl: A Survey of Municipal Laws Relating to Backyard Poultry and a Model Ordinance for Regulating City Chickens." *Environmental Law Reporter* (July 27, 2012): 888–920.

Eriksson, Jonas, Greger Larson, Ulrika Gunnarsson, et al. "Identification of the Yellow Skin Gene Reveals a Hybrid Origin of the Domestic Chicken." *PLoS Genetics* 4, no. 2 (February 29, 2008). https://doi.org/10.1371/journal.pgen.1000010.

Fabry, Merrill. "Now You Know: Which Came First, the Chicken or the Egg?" *Time*, September 21, 2016.

BIBLIOGRAPHY

Gerken, Sandie. "Sussex Home Grown Success Story: Cecile Steele's Pullet Surprise." *High Tide News* 4, no. 3 (March 2016): 1–2, 8.

Gibson, Kate. "The Biggest Source of Salmonella Outbreaks? It's Clucking in US Backyards." *CBS News*, July 30, 2019.

Grabell, Michael. "Exploitation and Abuse at the Chicken Plant." *New Yorker*, May 1, 2017.

Jacob, Jacquie. "Normal Behaviors of Chickens in Small and Backyard Poultry Flocks." Poultry Extension (website), May 5, 2015. https://poultry.extension.org/articles/poultrybehavior/normal-behaviors-of-chickens-in-small-and-backyard-poultry-flocks.

Jacourt, Louis. "Sacred Chickens." In *The Encyclopedia of Diderot and d'Alembert Collaborative Translation Project*. Translated by Dena Goodman. Ann Arbor: Michigan Publishing, University of Michigan Library, 2007. http://hdl.handle.net/2027/spo.did2222.0000.865.

Kravchenko, Julia, Sung Han Rhew, Igor Akushevich, Pankaj Agarwal, and H. Kim Lyerly. "Mortality and Health Outcomes in North Carolina Communities Located in Close Proximity to Hog Concentrated Animal Feeding Operations." *North Carolina Medical Journal* 79 (September 1, 2018): 278–88.

Larsen, Janet. "Peak Meat: US Meat Consumption Falling." Earth Policy Institute, March 7, 2012. www.earth-policy.org/data_highlights/2012/highlights25.

Loria, Joe. "Shocking ICE Raids Expose Horrific Conditions for Meat Industry Workers." Mercy for Animals, April 18, 2018. https://mercyforanimals.org/shocking-ice-raids-expose-horrific-conditions.

Løtvedt, Pia, Lejla Bektic, Amir Fallahshahroudi, and Jordi Altimiras Corderroure. "Chicken Domestication Changes Expression of Stress-Related Genes in Brain, Pituitary and Adrenals." *Neurobiology of Stress* 7 (2017): 113–21.

Moyer, Justin. "'You Need to Kill Him?': Tyson Food Contractors Caught on Video Mistreating Chickens." *Washington Post*, December 6, 2017.

National Center for Farmworker Health. "Poultry Workers" (fact sheet). 2014. www.ncfh.org/uploads/3/8/6/8/38685499/fs-poultryworkers.pdf.

National Chicken Council. "Chicken Consumption Continues to Soar in the US." July 11, 2016. https://www.nationalchickencouncil.org/chicken-consumption-continues-soar-u-s.

———. "Per Capita Consumption of Poultry and Livestock, 1965 to Estimated 2019, in Pounds." Accessed March 21, 2019. https://www.nationalchickencouncil.org/about-the-industry/statistics/per-capita

BIBLIOGRAPHY

-consumption-of-poultry-and-livestock-1965-to-estimated-2012-in-pounds.

Oxfam. "Keep Eating Chicken—But Ask for Some Justice on the Side." October 27, 2015. https://www.oxfamamerica.org/explore/stories/keep-eating-chickenbut-ask-for-some-justice-on-the-side.

Pimentel, David. "Soil Erosion: A Food and Environmental Threat." *Environment, Development and Sustainability* 8 (2006): 119–37.

Price, Edward O. "Behavioral Development in Animals Undergoing Domestication." *Applied Animal Behaviour Science* 65 (1999): 245–71.

Roth, Nataliya, Annemarie Käsbohrer, Singrid Mayrhofer, Ulrike Zitz, Charles Hofacre, and Konrad J. Domig. "The Application of Antibiotics in Broiler Production and Resulting Antibiotic Resistance in *Escherichia coli*: A Global Overview." *Poultry Science* 90, no. 4 (April 2019): 1791–1804.

Stuesse, Angela. "The Poultry Industry Recruited Them. Now ICE Raids Are Devastating Their Communities." *Washington Post*, August 9, 2019.

Tobin, Molly R., Jesse L. Goldshear, Lance B. Price, Jay P. Graham, and Jessica H. Leibler. "A Framework to Reduce Infectious Disease Risk from Urban Poultry in the United States." *Public Health Reports* 130, no. 4 (July 2015): 380–91.

US Government Accountability Office. "Workplace Safety and Health: Safety in the Meat and Poultry Industry, while Improving, Could Be Further Strengthened." January 2005. https://www.gao.gov/assets/250/245042.pdf.

Xiang, Hai, Jianqiang Gao, Baoquan Yu, et al. "Early Holocene Chicken Domestication in Northern China." *Proceedings of the National Academy of Sciences of the United States of America* 11, no. 49 (December 9, 2014): 17564–69.

CHAPTER 2: BIKING FOR THE BIRDS

American Veterinary Medical Association. "Welfare Implications of Beak Trimming." February 7, 2010. https://www.avma.org/resources-tools/literature-reviews/welfare-implications-beak-trimming#references.

Benson, G. John. "Pain in Farm Animals: Nature, Recognition, and Management." In *The Well-Being of Farm Animals: Challenges and Solutions*, edited by G. John Benson and Bernard E. Rollin, 61–84. Ames, IA: Blackwell, 2004.

Cheng, Heng-wei. "Laying Hen Welfare Fact Sheet: Current Developments in Beak-Trimming." US Department of Agriculture Livestock Behavior Research Unit. Fall 2010. www.ars.usda.gov/ARSUserFiles/50201500/Beak%20Trimming%20Fact%20Sheet.pdf.

BIBLIOGRAPHY

Digitale, Robert, and Susan Minichiello. "Dozens of Animal Welfare Activists Arrested after Large Protest at Petaluma Chicken Farm." *Press Democrat* (Santa Rosa, CA), May 29, 2018.

Direct Action Everywhere. "Following Mass Arrests at Petaluma Chicken Farm, Two Activists to be Arraigned on Seven Felony Charges Each." November 9, 2018. https://www.directactioneverywhere.com/thelibera tionist/2018/11/9/following-mass-arrests-at-petaluma-chicken-farm-two-activists-to-be-arraigned-on-felony-charges.

Food and Agriculture Organization of the United Nations. "Egg Production." In *Egg Marketing: A Guide for the Production and Sale of Eggs*. FAO Agricultural Services Bulletin 150. Rome: Food and Agriculture Organization of the United Nations, 2003, http://www.fao.org/3/Y4628E/y4628e03.htm#bm03.

Huang, J. C., N. Mu, J. Yang, M. Huang, and X. L. Xu. "Effects of Calcium Chloride on Calpain and Myofibril Protein Degradation of Spent Hen Breast Meat." In *63rd International Congress of Meat Science and Technology*, edited by Declan Troy, Clara McDonnell, Laura Hinds, and Joseph Kerry, 320–22. Wageningen, Netherlands: Wageningen Academic Publishers, 2017.

Jarvis, Erich D., Onur Güntürkün et. al. "Avian Brains and a New Understanding of Vertebrate Brain Evolution." *National Review of Neuroscience* 6, no. 2 (February 2005): 151–59.

Karsten, H. D., P. H. Patterson, R. Stout, and G. Crew. "Vitamins A, E and Fatty Acid Composition of the Eggs of Caged Hens and Pastured Hens." *Renewable Agriculture and Food Systems* 25, no 4 (January 12, 2010): 45–54.

Langley, Liz. "Chickens Prefer Attractive People." *National Geographic*, January 13, 2018.

Marino, Lori. "Thinking Chickens: A Review of Cognition, Emotion, and Behavior in the Domestic Chicken." *Animal Cognition* 20 (2017): 127–47.

Pacelle, Wayne. "Your Lunch Money: USDA Spends Millions on Spent Hens." *A Humane World: Kitty Block's Blog*. Humane Society of the United States, December 10, 2009. https://blog.humanesociety.org/2009/12/usda-hen-meat.html.

Shields, Sara, and Ian J. H. Duncan. "A Comparison of the Welfare of Hens in Battery Cages and Alternative Systems." *Impacts on Farm Animals* 18 (2009). https://animalstudiesrepository.org/hsus_reps_impacts_on_animals/18.

BIBLIOGRAPHY

Thomson, Julia R. "What Farms Do to Hens Who Are Too Old to Lay Eggs." *HuffPost*, May 14, 2018. https://www.huffpost.com/entry/egg-laying-hens_n_59c3c93fe4b0c90504fc04a1.

United Poultry Concerns. "Debeaking." UPC Fact Sheet, accessed April 30, 2020. https://www.upc-online.org/merchandise/debeak_factsheet.html.

US Department of Agriculture. *Poultry Industry Manual*. Ames: Center for Food Security and Public Health at Iowa State University of Science and Technology, in collaboration with US Department of Agriculture Animal and Plant Health Inspection Service, 2013.

CHAPTER 3: URBAN AGRICULTURE

Allcott, Hunt, Rebecca Diamond, Jean-Pierre Dubé, et al. "Food Deserts and the Causes of Nutritional Inequality." *Quarterly Journal of Economics* 134, no. 4 (2019): 1793–1844.

Brones, Anna. "Food Apartheid: The Root of the Problem with America's Groceries." *Guardian*, May 15, 2018.

Coyne, Kelly, and Erik Knutzen. *The Urban Homestead: Your Guide to Self-Sufficient Living in the Heart of the City*. Port Townsend, WA: Process Media, 2010.

De Bon, H., L. Parrot, and P. Moustier. "Sustainable Agriculture in Developing Countries: A Review." *Agronomy for Sustainable Development* 30, no. 1 (2010): 21–32.

Dettling, Lisa J., Joanne W. Hsu, Lindsay Jacobs, Kevin B. Moore, and Jeffrey P. Thompson with assistance from Elizabeth Llanes. "Recent Trends in Wealth-Holding by Race and Ethnicity: Evidence from the Survey of Consumer Finances." FEDS Notes. Washington DC: Board of Governors of the Federal Reserve System, September 27, 2017. https://doi.org/10.17016/2380-7172.2083.

Fess, Tiffany L., and Vagner A. Benedito. "Organic verses Conventional Cropping Sustainability: A Comparative Systems Analysis." *Sustainability* 10, no. 1 (2018). www.mdpi.com/journal/sustainability.

Finney, Carolyn. *Black Faces, White Spaces: Reimagining the Relationship of African Americans to the Great Outdoors*. Chapel Hill: University of North Carolina Press, 2014.

Huang, J. C., N. Mu, J. Yang, M. Huang, and X. L. Xu. "Effects of Calcium Chloride on Calpain and Myofibril Protein Degradation of Spent Hen Breast Meat." In *63rd International Congress of Meat Science and Technology*, edited by Declan Troy, Clara McDonnell, Laura Hinds, and Joseph

BIBLIOGRAPHY

Kerry, 320–22. Wageningen, Netherlands: Wageningen Academic Publishers, 2017.

Jennings, Angel. "Compton Gets Taste of Healthier Eating with New Farmers Market." *Los Angeles Times*, September 14, 2013.

Maeder, Paul, Andreas Fliessbach, David Dubois, Lucie Gunst, Padruot Fried, and Urs Niggli. "Soil Fertility and Biodiversity in Organic Farming." *Science* 296 (May 31, 2002): 1694–97.

Marino, Lori. "Thinking Chickens: A Review of Cognition, Emotion, and Behavior in the Domestic Chicken." *Animal Cognition* 20 (2017): 127–47.

McClintock, Nathan. "Why Farm the City? Theorizing Urban Agriculture through a Lens of Metabolic Rift." *Cambridge Journal of Regions, Economy, and Society* 3, no. 2 (July 2010): 191–207.

Morland, Kimberly, Steve Wing, Ana Diez Roux, and Charles Poole. "Neighborhood Characteristics Associated with the Location of Food Stores and Food Service Places." *American Journal of Preventive Medicine* 22, no. 1 (January 2002): 23–29.

Smit, J., A. Ratta, and J. Nasr. *Urban Agriculture: Food, Jobs, and Sustainable Cities*. New York: United Nations Development Programme, 1996.

Trust for Public Land of America. "Here's the Dirt on Park Trends: Community Gardens Are Growing." August 22, 2018. https://www.tpl.org/blog/here%E2%80%99s-dirt-park-trends-community-gardens-are-growing.

CHAPTER 4: A FREEGAN FLOCK

Agricultural Marketing Resource Center. "Organic Corn Profile." November 2018. https://www.agmrc.org/commodities-products/grains-oilseeds/corn-grain/organic-corn-profile.

Brucker, Monica. "Gulf of Mexico Dead Zone." Microbial Life Educational Resources, 2018. https://serc.carleton.edu/microbelife/topics/deadzone/index.html.

Environmental Defense Fund. "Methane: The Other Important Greenhouse Gas." Accessed May 2019. https://www.edf.org/climate/methane-other-important-greenhouse-gas.

Environmental Protection Agency. "Municipal Solid Waste." Last updated March 29, 2016. https://archive.epa.gov/epawaste/nonhaz/municipal/web/html.

Fischer, Bob, and Andy Lamey. "Field Deaths in Plant Agriculture." *Journal of Agriculture and Environmental Ethics* 31, no. 4 (August 2018): 409–28.

Foley, Jonathan. "It's Time to Rethink America's Corn System." *Scientific American*, March 5, 2013.

BIBLIOGRAPHY

Garcia, Valerie, Ellen Cooter, James Lewis Crooks, Brian Hinckley, Mark Murphy, and Xiangnan Xing. "Examining the Impacts of Increased Corn Production on Groundwater Quality Using a Coupled Modeling System." *Science of the Total Environment* 586 (May 15, 2017): 16–24.

Glover, Jerry D., and John P. Reganold. "Perennial Grains Food Security for the Future." *Issues in Science and Technology* 26, no. 2 (Winter 2010). https://issues.org/glover.

Klasing, Kirk C. "Nutritional Requirements of Poultry." *Merck Veterinary Manual.* Last modified May 2015. https://www.merckvetmanual.com/poultry/nutrition-and-management-poultry/nutritional-requirements-of-poultry#v4698325.

Kurutz, Steven. "Not Buying It." *New York Times*, June 21, 2007.

Milman, Oliver. "Americans Waste 150,000 Tons of Food Each Day—Equal to a Pound per Person." *Guardian*, April 18, 2018.

Pacelle, Wayne. "Your Lunch Money: USDA Spends Millions on Spent Hens." *A Humane World: Kitty Block's Blog.* The Humane Society of the United States, December 10, 2009. https://blog.humanesociety.org/2009/12/usda-hen-meat.html.

Ravindran, V. *Use of Cassava and Sweet Potatoes in Animal Feeding.* FAO Better Farming Series 46. Rome: Food and Agricultural Organization of the United Nations, 1995.

Solnit, Rebecca. "The Habits of Highly Cynical People." *Harper's*, May 2016.

Union of Concerned Scientists. "The Hidden Costs of Industrial Agriculture." August 24, 2008. https://www.ucsusa.org/food_and_agriculture/our-failing-food-system/industrial-agriculture/hidden-costs-of-industrial.html.

US Department of Agriculture. "Food Waste FAQs." Accessed April 31, 2020. https://www.usda.gov/foodwaste/faqs.

Zielinski, Sarah. "14 Fun Facts about Chickens." *Smithsonian*, August 31, 2011.

CHAPTER 5: PAMPERED POULTRY

Jha, Rajesh, Janelle M. Fouhse, Utsav P. Tiwari, Linge Li, and B. P. Wiling. "Dietary Fiber and Intestinal Health of Monogastric Animals." *Frontiers in Veterinary Science* 6, no. 48 (March 6, 2019): 1–11.

Mattocks, Jeff. "Pasture Raised Poultry Nutrition." Slide Share, prepared for Heifer International, November 17, 2002. https://www.slideshare.net/xx5v7/atx211.

Percy, Pam. *The Field Guide to Chickens.* St. Paul, MN: Voyageur Press, 2006.

BIBLIOGRAPHY

Spencer, Terrell. "Pastured Poultry Nutrition and Forages." National Sustainable Agriculture Information Service, August 2013. https://attra3.ncat.org/attra-pub-summaries/?pub=452.

CHAPTER 6: FOWL FEAST

American Meat Association. "Salmonella Fact Sheet." 2015. https://meatscience.org/docs/default-source/publications-resources/fact-sheets/salmonella-fact-sheet-2015.pdf?sfvrsn=87518eb3_0.

Arias, Cesar A., and Jose M. Munita. "Mechanisms of Antibiotic Resistance." *Microbiology Spectrum* 4, no. 2 (April 2016). https://doi.org/10.1128/microbiolspec.VMBF-0016-2015.

Ball-Blakely, Christine. "CAFOs: Plaguing North Carolina Communities of Color." *Sustainable Department Law and Policy* 18, no. 1 (Fall 2017): 4–49.

Brulliard, Karin. "Kissing Chickens Is Bad for Your Health, CDC Warns." *Washington Post*, September 14, 2016.

California Department of Food and Agriculture. "What You Need to Know: Virulent Newcastle Disease." May 2019. https://www.cdfa.ca.gov/ahfss/Animal_Health/pdfs/FAQ-WhatYouNeedtoKnow-vND2018-19.pdf.

Center for Infectious Disease Research and Policy. "Exotic Newcastle Disease Battle Ends." September 17, 2003. https://www.cidrap.umn.edu/news-perspective/2003/09/exotic-newcastle-disease-battle-ends.

Centers for Disease Control and Prevention. "CDC Investigation Notice: Multidrug-Resistant Salmonella Outbreak Linked to Raw Chicken Products." February 21, 2019. https://www.cdc.gov/media/releases/2019/s0221-chicken-salmonella-outbreak.html.

———. "Don't Play Chicken with Your Health." June 3, 2019. https://www.cdc.gov/healthypets/resources/dont-play-chicken-with-your-health-P.pdf.

———. "Multiple Outbreaks of *Salmonella* Infections Linked to Contact with Live Poultry in Backyard Flocks, 2018." September 13, 2018. https://www.cdc.gov/salmonella/backyard-flocks-06-18/index.html.

———. "Salmonella: Previous Outbreaks." 2019. https://www.cdc.gov/salmonella/outbreaks.html.

———. "Surveillance for Foodborne Disease Outbreaks—United States, 2009–2015." July 27, 2018. https://www.cdc.gov/mmwr/volumes/67/ss/ss6710a1.htm.

Chicken Run Rescue. "Chicken Care Certification." 2019. www.chickenrunrescue.org/Chicken-Care-Certification.

BIBLIOGRAPHY

Christensen, Jen, and Debra Goldschmidt. "Why Backyard Chickens Are a Health Risk." CNN, July 26, 2018. https://www.cnn.com/2018/07/26/health/backyard-chickens-salmonella/index.html.

Conniff, Richard. "The Man Who Turned Antibiotics into Animal Feed." Alicia Patterson Foundation, September 9, 2012. https://aliciapatterson.org/stories/man-who-turned-antibiotics-animal-feed.

Fassler, Joe. "CDC: Backyard Chickens are Giving Their Well-Meaning Owners Salmonella." The Counter, August 22, 2017. https://thecounter.org/backyard-chickens-salmonella-safety/.

Food Empowerment Project. "Environmental Racism." Accessed September 19, 2019. https://foodispower.org/environmental-and-global/environmental-racism.

Humane Society of the United States. "Adopting and Caring for Backyard Chickens." April 2019. https://www.humanesociety.org/resources/adopting-and-caring-backyard-chickens.

"Many Chicken Farms in Chicago Slums." *Chicago Daily Tribune*, May 9, 1909, p. H2.

Marshall, Bonnie M., and Stuart B. Levy. "Food Animals and Antimicrobials: Impacts on Human Health." *Clinical Microbiology Review* 24, no. 4 (October 2011):718–33.

Martin, Michael J., Sapna E. Thottathil, and Thomas B. Newman. "Antibiotics Overuse in Animal Agriculture: A Call to Action for Health Care Providers." *American Journal of Public Health* 105, no. 12 (December 2015): 2409–10.

May, Ashley. "Over 200 Salmonella Infections Linked to Backyard Chickens, CDC Warns." *USA Today*, July 25, 2018.

McKenna, Maryn. *Big Chicken: The Incredible Story of How Antibiotics Created Modern Agriculture and Changed the Way the World Eats*. Washington, DC: National Geographic, 2017.

McWilliams, James. "Five Reasons Why Owning Backyard Chickens Is for the Birds." *Forbes*, November 21, 2013.

National Research Council (US) Committee to Study the Human Health Effects of Subtherapeutic Antibiotic Use in Animal Feeds. "Appendix K: Antibiotics in Animal Feeds." In *The Effects of Human Health of Subtherapeutic Use of Antimicrobials in Animal Feeds*, 317–76. Washington, DC: National Academies Press, 1980.

Ogle, Maureen. "Riots, Rage, and Resistance: A Brief History of How Antibiotics Arrived on the Farm." *Scientific American* (blog), September 3,

BIBLIOGRAPHY

2013. https://blogs.scientificamerican.com/guest-blog/riots-rage-and-resistance-a-brief-history-of-how-antibiotics-arrived-on-the-farm.

Palmer, Kim. "Backyard Chicken Trend Comes Home to Roost." *Minneapolis Star Tribune*, October 8, 2013.

Pattani, Aneri. "Backyard Chickens Carry a Hidden Risk: Salmonella." *New York Times*, September 4, 2017.

Raffensperger, Carolyn, and Joel Tickner. *Protecting Public Health and the Environment: Implementing the Precautionary Principle*. Washington, DC: Island Press, 1999.

US Department of Agriculture, Animal and Plant Health Inspection Services. "Changes to the Salmonella and Campylobacter Verification Testing Program." January 26, 2015. https://www.federalregister.gov/documents/2015/01/26/2015-01323/changes-to-the-salmonella-and-campylobacter-verification-testing-program-proposed-performance.

———. "Salmonella Verification Testing Program Monthly Posting." Referencing May 27, 2018 to May 25, 2019. https://www.fsis.usda.gov/wps/portal/fsis/topics/data-collection-and-reports/microbiology/salmonella-verification-testing-program.

———. "USDA Confirms Virulent Newcastle Disease in Pet Chickens in Arizona: Not a Food Safety Concern." April 5, 2019. https://www.aphis.usda.gov/aphis/newsroom/stakeholder-info/sa_by_date/2019/sa-04/vnd-arizona.

World Health Organization. "At the UN, Global Leaders Commit to Act on Antimicrobial Resistance." September 21, 2016. https://www.who.int/news-room/detail/21-09-2016-at-un-global-leaders-commit-to-act-on-antimicrobial-resistance.

CHAPTER 7: EATING BUGS FOR THE ENVIRONMENT

Lappé, Frances Moore. *Diet for a Small Planet*. 20th Anniversary Edition. New York: Ballantine Books, 1991.

Oaklander, Mandy, and Lon Tweeten. "Should You Be Eating Bugs Instead of Meat?" *Time*, April 14, 2016.

Onsongo, V. O., M. Osuga, C. K. Gachuiri, et al. "Insects for Income Generation through Animal Feed: Effects of Dietary Replacement of Soybean and Fish Meal with Black Soldier Fly Meal on Broiler Growth and Economic Performance." *Journal of Economic Entomology* 111, no. 4 (August 2018):1966–73.

Shumo, Marwa, Issac M. Osuga, Fathiya M. Khamis, et al. "The Nutritive Value of Black Soldier Fly Larvae Reared on Common Organic Waste

BIBLIOGRAPHY

Streams in Kenya." *Scientific Reports* 9, no. 1 (July 2019). https://doi.org/10.1038/s41598-019-46603-z.

Smil, Vaclav. "Eating Meat: Evolution, Patterns, and Consequences." *Population and Development Review* 28, no. 4 (December 2002): 599–639.

van Huis, Arnold, Joost Van Itterbeeck, Harmke Klunder, Esther Mertens, Afton Halloran, Giulia Muir, and Paul Vantomme. "Edible Insects: Future Prospects for Food and Feed Security." FAO Forestry Paper 171. Rome: Food and Agriculture Organization of the United Nations, 2013.

CHAPTER 8: PRODUCTIVE PETS

American Society for the Prevention of Cruelty toward Animals (ASPCA). "A Growing Problem: Selective Breeding in the Chicken Industry: The Case for Slower Growth." November 2015. https://www.aspca.org/sites/default/files/chix_white_paper_nov2015_lores.pdf.

Bugos, Glenn E. "Intellectual Property Protection in the American Chicken-Breeding Industry." *Business History Review* 66, no. 1 (1992):127–68.

Canland, Douglas K., and Daniel H. Conklyn. "Social Dominance and Learning in the Domestic Chicken." *Psychonomic Science* 11, no. 7 (1968): 247–48.

Collins, N. E. "The Development of Social Behavior in Birds." *Auk: A Quarterly Journal of Ornithology* 69, no. 2 (April 1952): 127–59.

Davis, Karen. "Eliminating the Suffering of Chickens Bred for Meat." One Green Planet, 2012. https://www.onegreenplanet.org/animalsandnature/eliminating-the-suffering-of-chickens-bred-for-meat-2.

Farm Sanctuary. "Chicken Care: Large Breed." Accessed March 30, 2020. https://www.farmsanctuary.org/wp-content/uploads/2012/06/Animal-Care-Large-Breed-Chicken.pdf.

Hargis, Billy M. "Ascites Syndrome in Poultry." *Merck Veterinary Manual*. Last modified October 2014. https://www.merckvetmanual.com/poultry/miscellaneous-conditions-of-poultry/ascites-syndrome-in-poultry.

Humane Society of the United States. "An HSUS Report: The Welfare of Animals in the Chicken Industry." December 2013. https://www.humanesociety.org/sites/default/files/docs/hsus-report-welfare-chicken-industry.pdf.

Jacob, Jacquie. "Normal Behaviors of Chickens in Small and Backyard Poultry Flocks." Poultry Extension website, May 5, 2015. https://poultry.extension.org/articles/poultrybehavior/normal-behaviors-of-chickens-in-small-and-backyard-poultry-flocks.

BIBLIOGRAPHY

Kettlewell, P. J., and M. A. Mitchell. "Catching, Handling and Loading of Poultry for Road Transportation." *World's Poultry Science Journal* 50 (March 1994): 54–56.

Knowles, Toby G., Steve C. Kestin, Susan M. Haslam, et al. "Leg Disorders in Broiler Chickens: Prevelence, Risk Factors, and Prevention." *PloS ONE* 3, no. 2 (February 6, 2008). https://doi.org/10.1371/journal.pone.0001545.

Lange, Karen. "Super-Size Problem." Humane Society of the United States, March 1, 2017. https://www.humanesociety.org/news/super-size-problem-broiler-chickens?credit=web_id66423299.

Massachusetts Society for the Prevention of Cruelty to Animals. "Farm Animal Welfare: Chickens." Accessed April 31, 2020. https://www.mspca.org/animal_protection/farm-animal-welfare-chickens.

McKenna, Maryn. "The Surprising Origin of Chicken as a Dietary Staple." *National Geographic*, May 1, 2018.

North, Mack O., and Donald D. Bell. *Commercial Chicken Production Manual*. 4th ed. New York: Springer, 1990.

Price, John. "A Remembrance of Thorleif Schjelderup-Ebbe." *Human Ethology Bulletin* 10, no. 1 (March 1995): 1–6.

Ritz, Casey, M. Czarick, and Amanda Webster. "Evaluation of Hot Weather Thermal Environment and Incidence of Mortality Associated with Broiler Live Haul." *Journal of Applied Poultry Research* 14, no. 3 (September 2015): 594–602.

US Department of Agriculture. Food Availability Spreadsheet Recording.

Wallner-Pendleton, Eva, and Sheila Scheideler. "NF95-216 Ascites Syndrome in Broiler Chickens." Historical Materials from University of Nebraska–Lincoln Extension, 1995. https://digitalcommons.unl.edu/cgi/viewcontent.cgi?referer=&httpsredir=1&article=1149&context=extensionhist.

Wiehoff, Dale. "How the Chicken of Tomorrow Became the Chicken of the World." Institute for Agricultural and Trade Policy, March 26, 2013. https://www.iatp.org/blog/201303/how-the-chicken-of-tomorrow-became-the-chicken-of-the-world.

Wilkie, Rhoda. "Sentient Commodities and Productive Paradoxes: The Ambiguous Nature of Human-Livestock Relations in Northwest Scotland." *Journal of Rural Studies* 21, no. 2 (2005): 213–30.

CHAPTER 9: SLAUGHTERHOUSE IN THE BACKYARD

Adams, Carol. *The Sexual Politics of Meat: A Feminist-Vegetarian Critical Theory*. New York: Bloomsbury Academic, 2017.

BIBLIOGRAPHY

Berry, Wendell. "The Pleasures of Eating." In *What Are People For? Essays*. San Francisco: North Point Press, 1990.

Carrington, Damian. "Eating Less Meat Essential to Curb Climate Change, Says Report." Our World, December 5, 2014. https://ourworld.unu.edu/en/eating-less-meat-essential-to-curb-climate-change-says-report.

Davis, Karen. "The Need for Legislation and Elimination of Electrical Immobilization." United Poultry Concerns. Accessed May 2019. https://www.upc-online.org/slaughter/report.html.

Milo, Ron, Yinon Bar-On, and Rob Phillips. "The Biomass Distribution on Earth." *Proceedings of the National Academy of Sciences* 115, no. 2 (June 19, 2018): 6505–11.

CHAPTER 10: WASTE NOT, WANT NOT

Allen, Kristi. "Why China and America Fight over Chicken Feet." Atlas Obscura, January 28, 2019. https://www.atlasobscura.com/articles/chicken-feet-trade.

Bah, Clara S. F., Alaa El-Din A. Bekhit, Alan Carne, and Michelle A. McConnell. "Slaughterhouse Blood: An Emerging Source of Bioactive Compounds." *Comprehensive Reviews in Food Science and Food Safety* 12, no. 3 (May 2013): 314–31.

Child, Julia. *Mastering the Art of French Cooking*. New York: Knopf, 1964.

Cross, Sue. "Slaughterhouse Waste—It Has to Be Dealt With: A Case for the Vegan Option Continued." *HuffPost*, February 14, 2013. https://www.huffingtonpost.co.uk/sue-cross/horse-meat-slaugtherhorse-veganism_b_2684502.html.

Flock, Eunice, Jesse L. Bollman, Harold R. Hester, and Frank C. Mann. "Fatty Livers in the Goose Produced by Overfeeding." *Journal of Biological Chemistry* 121, no. 1 (1937):117–29.

Gonzáles-Domíguez, Raúl, Tamara García-Barrera, and José Luis Gómez-Ariza. "Homeostasis of Metals in the Progression of Alzheimer's Disease." *BioMetals* 27 (2014): 539–49.

Groeger, Lena V. "And You Thought It Was Just 'Pink' Slime." *ProPublica*, April 12, 2012. https://www.propublica.org/article/and-you-thought-it-was-just-pink-slime.

Guémené, Daniel, and G. Guy. "The Past, Present and Future of Force-Feeding and 'Foie Gras' Production." *World's Poultry Science Journal* 60, no. 2 (June 2004): 210–22.

Henderson, Fergus. *The Whole Beast: Nose to Tail Eating*. New York: CCC, 2004.

BIBLIOGRAPHY

Krauss, Clifford. "Chewy Chicken Feet May Quash a Trade War." *New York Times*, September 16, 2009.

Lynch, Sarah A., Anne Maria Mullen, Eileen E. O'Neill, and Carlos Álvarez García. "Harnessing the Potential of Blood Proteins as Functional Ingredients: A Review of the State of the Art in Blood Processing." *Comprehensive Reviews in Food Science and Food Safety* 16, no. 2 (March 2017): 330–44.

Paul, Elisabeth. "Blood and Egg." *Nordic Food Lab*, January 7, 2014. https://futureconsumerlab.ku.dk/lab-facilities/nordic-food-lab-archive/Nordic_Lab_FINAL_2020.pdf.

Saint-Germain, Claire. "The Production of Bone Broth: A Study in Nutritional Exploitation." *Anthropozoologica* 25, no. 26 (1997): 153–56.

US Department of Agriculture, Food Safety and Inspection Service. "Giblets and Food Safety." September 2008. https://www.fsis.usda.gov/wps/wcm/connect/8c532492-e9a6-43c7-abff-d5a00cb3f642/Giblets_and_Food_Safety.pdf?MOD=AJPERES.

CHAPTER 11: AFTER HARVEST

Bajželj, Bojana, Keith S. Richards, Julian M. Allwood, Pete Smith, John S. Dennis, Elizabeth Curmi, and Christopher A. Gilligan. "Importance of Food-Demand Management for Climate Mitigation." *Nature Climate Change* 4 (2014): 924–29.

Carrington, Damian. "Eating Less Meat Essential to Curb Climate Change, Says Report." Our World, December 5, 2014. https://ourworld.unu.edu/en/eating-less-meat-essential-to-curb-climate-change-says-report.

Heggie, Jon. "The Future of Livestock Farming." *National Geographic*, March 18, 2019.

ABOUT THE AUTHOR

Gina G. Warren grew up in California, where she was introduced to chickens as a child. Her parents adopted hens from a neighbor, and within weeks Gina and her friends were walking leashed chickens to the park. Since then, Gina has studied English and philosophy at Pacific University in Oregon, received an MFA from the Northwest Institute of Literary Arts in Washington, and obtained a PhD in English from the University of Louisiana at Lafayette. Her creative nonfiction has appeared in publications including *Entropy, Orion, Creative Nonfiction,* and *Terrain.org.*